LONDON MATHEMATICAL SOCIETY LECTURE NOTE SERIES

Managing Editor: Professor M. Reid, Mathematics Institute,
University of Warwick, Coventry CV4 7AL, United Kingdom

The titles below are available from booksellers, or from Cambridge University Press at
http://www.cambridge.org/mathematics

London Mathematical Society Lecture Note Series: 432

Topics in Graph Automorphisms and Reconstruction

Second Edition

JOSEF LAURI
University of Malta

RAFFAELE SCAPELLATO
Politecnico di Milano

CAMBRIDGE
UNIVERSITY PRESS

CAMBRIDGE
UNIVERSITY PRESS

Shaftesbury Road, Cambridge CB2 8EA, United Kingdom

One Liberty Plaza, 20th Floor, New York, NY 10006, USA

477 Williamstown Road, Port Melbourne, VIC 3207, Australia

314–321, 3rd Floor, Plot 3, Splendor Forum, Jasola District Centre, New Delhi – 110025, India

103 Penang Road, #05–06/07, Visioncrest Commercial, Singapore 238467

Cambridge University Press is part of Cambridge University Press & Assessment,
a department of the University of Cambridge.

We share the University's mission to contribute to society through the pursuit of
education, learning and research at the highest international levels of excellence.

www.cambridge.org
Information on this title: www.cambridge.org/9781316610442

First published 2016

A catalogue record for this publication is available from the British Library

Library of Congress Cataloging-in-Publication data
Names: Lauri, Josef, 1955– | Scapellato, Raffaele, 1955–
Title: Topics in graph automorphisms and reconstruction /
Josef Lauri and Raffaele Scapellato.
Description: 2nd edition. | Cambridge : Cambridge University Press, 2016. |
Series: London Mathematical Society lecture note series; 432 |
Includes bibliographical references and index.
Identifiers: LCCN 2016014849 | ISBN 9781316610442 (pbk. : alk. paper)
Subjects: LCSH: Graph theory. | Automorphisms. | Reconstruction (Graph theory)
Classification: LCC QA166.L39 2016 | DDC 511/.5–dc23
LC record available at https://lccn.loc.gov/2016014849

ISBN   978-1-316-61044-2   Paperback

Lil
Mary Anne,
Christina, Beppe u Sandrina

A
Fiorella

Contents

Preface to the Second Edition

In this second edition of our book we have tried to maintain the same structure as the first edition, namely a text which, although not providing an exhaustive coverage of graph symmetries and reconstruction, provides a detailed coverage of some particular areas (generally motivated by our own research interest), which is not a haphazard collection of results but which presents a clear pathway through this thick forest. And our aim remains that of producing a text which can relatively quickly guide the reader to the point of being able to understand and carry out research in the topics which we cover.

Among the additions in this edition we point out the use of the free computer programs GAP, GRAPE and *Sage* to construct and investigate some well-known graphs, including examples with properties like being semisymmetric, a topic which was treated in the first edition but for which examples are not easy to construct 'by hand'. We have also updated some chapters with new results, improved the presentation and proofs of others, and introduced short treatments of topics such as character theory of abelian groups and their Cayley graphs to emphasise the connection between graph theory and other areas of mathematics.

We have corrected a number of errors which we found in the first edition, and for this we would like to thank colleagues who have pointed out several of them, particularly Bill Kocay, Virgilio Pannone and Alex Scott.

A special thanks goes to Russell Mizzi for help with overhauling Chapter 6, where we also introduce the new idea of two-fold isomorphisms, and to Leonard Soicher and Matan Zif-Av for several helpful tips regarding the use of GAP and GRAPE.

The second author would like to thank the Politecnico di Milano for giving him the opportunity, by means of a sabbatical, to focus on the work needed to complete the current edition of this book. He also thanks the University of Malta for its kind hospitality during this sabbatical.

The authors will maintain a list of corrections and addenda at http://staff.um .edu.mt/josef.lauri.

Josef Lauri
Raffaele Scapellato

Preface to the First Edition

This book arose out of lectures given by the first author to Masters students at the University of Malta and by the second author at the Università Cattolica di Brescia.

This book is not intended to be an exhaustive coverage of graph theory. There are many excellent texts that do this, some of which are mentioned in the References. Rather, the intention is to provide the reader with a more in-depth coverage of some particular areas of graph theory. The choice of these areas has been largely governed by the research interests of the authors, and the flavour of the topics covered is predominantly algebraic, with emphasis on symmetry properties of graphs. Thus, standard topics such as the automorphism group of a graph, Frucht's Theorem, Cayley graphs and coset graphs, and orbital graphs are presented early on because they provide the background for most of the work presented in later chapters. Here, more specialised topics are tackled, such as graphical regular representations, pseudosimilarity, graph products, Hamiltonicity of Cayley graphs and special types of vertex-transitive graphs, including a brief treatment of the difficult topic of classifying vertex-transitive graphs. The last four chapters are devoted to the Reconstruction Problem, and even here greater emphasis is given to those results that are of a more algebraic character and involve the symmetry of graphs. A special chapter is devoted to graph products. Such operations are often used to provide new examples from existing ones but are seldom studied for their intrinsic value.

Throughout we have tried to present results and proofs, many of which are not usually found in textbooks but have to be looked for in journal papers. Also, we have tried, where possible, to give a treatment of some of these topics that is different from the standard published material (for example, the chapter on graph products and much of the work on reconstruction).

xiv   *Preface to the First Edition*

Although the prerequisites for reading this book are quite modest (exposure to a first course in graph theory and some discrete mathematics, and elementary knowledge about permutation groups and some linear algebra), it was our intention when preparing this book that a student who has mastered its contents would be in a good position to understand the current state of research in most of the specialised topics covered, would be able to read with profit journal papers in these areas, and would hopefully have his or her interest sufficiently aroused to consider carrying out research in one of these areas of graph theory.

We would finally like to thank Professor Caroline Series for showing an interest in this book when it was still in an early draft form and the staff at Cambridge University Press for their help and encouragement, especially Roger Astley, Senior Editor, Mathematical Sciences, and, for technical help with LaTeX, Alison Woollatt, who, with a short style file, solved problems that would have baffled us for ages. Thanks are also due to Elise Oranges, who edited this book thoroughly and pointed out several corrections.

The first author would also like to thank the Academic Work Resources Fund Committee and the Computing Services Centre of the University of Malta, the first for some financial help while writing this book and the second for technical assistance. He also thanks his M.Sc. students at the University of Malta, who worked through draft chapters of this book and whose comments and criticism helped to improve the final product.

Josef Lauri
Raffaele Scapellato

1

Graphs and Groups: Preliminaries

1.1 Graphs and digraphs

In these chapters a *graph* $G = (V(G), E(G))$ will consist of two disjoint sets: a nonempty set $V = V(G)$ whose elements will be called *vertices* and a set $E = E(G)$ whose elements, called *edges*, will be unordered pairs of distinct elements of V. Unless explicitly stated otherwise, the set of vertices will always be finite. An edge, $\{u, v\}$, $u, v \in V$, is also denoted by uv. Sometimes E is allowed to be a multiset, that is, the same edge can be repeated more than once in E. Such edges are called *multiple edges*. Also, edges uu consisting of a pair of repeated vertices are sometimes allowed; such edges are called *loops*. But unless otherwise stated, it will always be assumed that a graph does not have loops or multiple edges. The *complement* of the graph G, denoted by $\overline{G}$, has the same vertex-set as G, but two distinct vertices are adjacent in the complement if and only if they are not adjacent in G.

The *degree* of a vertex v, denoted by $\deg(v)$, is the number of edges in $E(G)$ to which v belongs. A vertex of degree k is sometimes said to be a *k-vertex*. Two vertices belonging to the same edge are said to be *adjacent*, while a vertex and an edge to which it belongs are said to be *incident*. A loop incident to a vertex v contributes a value of 2 to $\deg(v)$. A graph is said to be *regular* if all of its vertices have the same degree. A regular graph with degree equal to 3 is sometimes called *cubic*. The minimum and maximum degrees of G are denoted by $\delta = \delta(G)$ and $\Delta = \Delta(G)$, respectively.

In general, given any two sets A, B, then $A - B$ will denote their set-theoretical difference, that is, the set consisting of all of the elements that are in A but not in B. Also, a set containing k elements is often said to be a *k-set*.

If S is a set of vertices of a graph G, then $G - S$ will denote the graph obtained by removing S from $V(G)$ and removing from $E(G)$ all edges incident to some vertex in S. If F is a set of edges of G, then $G - F$ will denote the graph whose

1

vertex-set is $V(G)$ and whose edge-set is $E(G) - F$. If $S = \{u\}$ and $F = \{e\}$, we shall, for short, denote $G - S$ and $G - F$ by $G - u$ and $G - e$, respectively.

If S is a subset of the vertices of G, then $G[S]$ will denote the subgraph of G *induced* by S, that is, the subgraph consisting of the vertices in S and all of the edges joining pairs of vertices from S.

An important modification of the foregoing definition of a graph gives what is called a *directed graph*, or *digraph* for short. In a digraph $D = (V(D), A(D))$ the set $A = A(D)$ consists of ordered pairs of vertices from $V = V(D)$ and its elements are called *arcs*. Again, an arc (u, v) is sometimes denoted by uv when it is clear from the context whether we are referring to an arc or an edge. The arc uv is said to be *incident to v* and *incident from u*; the vertex u is said to be *adjacent to v* whereas v is *adjacent from u*. The number of arcs incident from a vertex v is called its *out-degree*, denoted by $\deg_{out}(v)$, while the number of arcs incident to v is called its *in-degree* and is denoted by $\deg_{in}(v)$. A digraph is said to be *regular* if all of its vertices have the same out-degree or, equivalently, the same in-degree. Sometimes, when we need to emphasise the fact that a graph is not directed, we say that it is *undirected*.

The number of vertices of a graph G or digraph D is called its *order* and is generally denoted by $n = n(G)$ or $n = n(D)$, while the number of edges or arcs is called its *size* and is denoted by $m = m(G)$ or $m = m(D)$.

A sequence of distinct vertices of a graph, $v_1, v_2, \ldots, v_{k+1}$, and edges $e_1, e_2, \ldots, e_k$ such that each edge $e_i = v_i v_{i+1}$ is called a *path*. If we allow v_1 and v_{k+1}, and only those, to be the same vertex, then we get what is called a *cycle*.

The *length* of a path or a cycle in G is the number of edges in the path or cycle. A path of length k is denoted by P_{k+1} while a cycle of length k is denoted by C_k. The *distance* between two vertices u, v in a connected graph G, denoted by $d(u, v)$, is the length of the shortest path joining u and v. The *diameter* of G is the maximum value attained by $d(u, v)$ as u, v run over $V(G)$, and the *girth* is the length of the shortest cycle.

In these definitions, if we are dealing with a digraph and the $e_i = v_i v_{i+1}$ are arcs, then the path or cycle is called a *directed path* or *directed cycle*, respectively.

Given a digraph D, the *underlying graph* of D is the graph obtained from D by considering each pair in $A(D)$ to be an unordered pair. Given a graph G, the digraph $\overleftrightarrow{G}$ is obtained from G by replacing each edge in $E(G)$ by a pair of oppositely directed arcs. This way, a graph can always be seen as a special case of a digraph.

We adopt the usual convention of representing graphs and digraphs by drawings in which each vertex is shown by a dot, each edge by a curve joining the corresponding pair of dots and each arc (u, v) by a curve with an arrowhead pointing in the direction from u to v.

A number of definitions on graphs and digraphs will be given as they are required. However, several standard graph theoretic terms will be used but not defined in these chapters; these can be found in any of the references [257] or [259].

1.2 Groups

A *permutation group* will be a pair (Γ, Y) where Y is a finite set and Γ is a subgroup of the symmetric group S_Y, that is, the group of all permutations of Y. The stabiliser of an element $y \in Y$ under the action of Γ is denoted by Γ_y while the orbit of y is denoted by $\Gamma(y)$. The *Orbit-Stabiliser Theorem* states that, for any element $y \in Y$,

$$|\Gamma| = |\Gamma(y)| \cdot |\Gamma_y|.$$

If the elements of Y are all in one orbit, then (Γ, Y) is said to be a *transitive permutation group* and Γ is said to act *transitively* on Y. The permutation group Γ is said to act *regularly* on Y if it acts transitively and the stabiliser of any element of Y is trivial. By the Orbit-Stabiliser Theorem, this is equivalent to saying that Γ acts transitively on Y and $|\Gamma| = |Y|$. Also, Γ acts regularly on Y is equivalent to saying that, for any $y_1, y_2 \in Y$, there exists exactly one $\alpha \in \Gamma$ such that $\alpha(y_1) = y_2$.

One important regular action of a permutation group arises as follows. Let Γ be any group, let $Y = \Gamma$ and, for any $\alpha \in \Gamma$, let λ_α be the permutation of Y defined by $\lambda_\alpha(\beta) = \alpha\beta$. Let $L(\Gamma)$ be the set of all permutations λ_α for all $\alpha \in \Gamma$. Then $(L(\Gamma), Y)$ defines a permutation group acting regularly on Y. This is called the *left regular representation* of the group Γ on itself. One can similarly consider the *right regular representation* of the group Γ on itself, and this is denoted by $(R(\Gamma), Y)$.

The following is an important generalisation of the previous definitions. If Γ is a group and $\mathcal{H} \leq \Gamma$, let $Y = \Gamma/\mathcal{H}$ be the set of left cosets of $\mathcal{H}$ in Γ. For any $\alpha \in \Gamma$, let $\lambda_\alpha^{\mathcal{H}}$ be a permutation on Y defined by $\lambda_\alpha^{\mathcal{H}}(\beta\mathcal{H}) = \alpha\beta\mathcal{H}$. Let $L^{\mathcal{H}}(\Gamma)$ be the set of all $\lambda_\alpha^{\mathcal{H}}$ for all $\alpha \in \Gamma$. Then $(L^{\mathcal{H}}(\Gamma), Y)$ defines a permutation group that reduces to the left regular representation of Γ if $\mathcal{H} = \{1\}$.

Two permutation groups $(\Gamma_1, Y_1), (\Gamma_2, Y_2)$ are said to be *equivalent*, denoted by $(\Gamma_1, Y_1) \equiv (\Gamma_2, Y_2)$, if there exists a bijective isomorphism $\phi : \Gamma_1 \to \Gamma_2$ and a bijection $f : Y_1 \to Y_2$ such that, for all $y \in Y_1$ and for all $\alpha \in \Gamma_1$,

$$f(\alpha(x)) = \phi(\alpha)(f(x)).$$

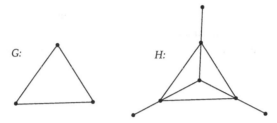

Figure 1.1. Aut(G), Aut(H) are isomorphic but not equivalent

In this case we also say that the action of Γ_1 on Y_1 is equivalent to the action of Γ_2 on Y_2, and sometimes we denote this simply by $\Gamma_1 = \Gamma_2$, when the two sets on which the groups are acting is clear from the context.

Figure 1.1 shows a simple example of two graphs whose automorphism groups (to be defined later in this chapter) are isomorphic as abstract groups but clearly not equivalent as permutation groups since the sets (of vertices) on which they act are not equal. (See also Exercise 1.7.)

Note in particular that, if $(\Gamma_1, Y_1) \equiv (\Gamma_2, Y_2)$, then apart from $\Gamma_1 \simeq \Gamma_2$ as abstract groups, and $|Y_1| = |Y_2|$, the cycle structure of the permutations of Γ_1 on Y_1 must be the same as those of Γ_2 on Y_2. However, the converse is not true; that is, Γ_1 and Γ_2 could be isomorphic and the cycle structures of their respective actions could be the same, but (Γ_1, Y_1) might not be equivalent to (Γ_2, Y_2) (see Exercise 1.9).

If (Γ, Y) is a permutation group acting on Y and Y' is a union of orbits of Y, then we can talk about the action of Γ *restricted* to Y', that is, the permutation group (Γ, Y') where, for $\alpha \in \Gamma$ and $y' \in Y'$, $\alpha(y')$ is the same as in (Γ, Y). When Y' is a union of orbits we also say that it is *invariant* under the action of Γ because in this case $\alpha(y') \in Y'$ for all $\alpha \in \Gamma$ and $y' \in Y'$. Also, (Γ', Y') is said to be a subpermutation group of (Γ, Y) if $\Gamma' \leq \Gamma$ and Y' is a union of orbits of Γ' acting on Y.

The following is a useful well-known result on permutation groups whose proof is not difficult and is left as an exercise (see Exercise 1.10).

Theorem 1.1 *Let (Γ, Y) be a permutation group acting transitively on Y. Let $y \in Y$, let $\mathcal{H} = \Gamma_y$ be the stabiliser of y and let W be $\Gamma/\mathcal{H}$, the set of left cosets of $\mathcal{H}$ in Γ. Then (Γ, Y) is equivalent to $(L^{\mathcal{H}}(\Gamma), W)$.*

If (Γ, Y) is not transitive, and $\mathcal{O}$ is the orbit containing y, then $(L^{\mathcal{H}}(\Gamma), W)$ is equivalent to the action of Γ on Y restricted to $\mathcal{O}$.

In the context of groups and graphs we shall need the very important idea of a group acting on pairs of elements of a set. Thus, let (Γ, Y) be a permutation group acting on the set Y. By $(\Gamma, Y \times Y)$ we shall mean the action on ordered pairs of Y induced by Γ as follows: If $\alpha \in \Gamma$ and $x, y \in Y$, then

$$\alpha((x, y)) = (\alpha(x), \alpha(y)).$$

Similarly, by $(\Gamma, \binom{Y}{2})$ we shall mean the action on unordered pairs of distinct elements of Y induced by

$$\alpha(\{x, y\}) = \{\alpha(x), \alpha(y)\}.$$

These ideas will be developed further in a later chapter.

In later chapters we shall also need the notions of k-transitivity and primitivity of a permutation group. In order to study permutation groups in more detail one has to dig deeper into the concept of transitivity. Suppose, for example, that Y is the set $\{1, 2, 3, 4, 5\}$ and Γ is the group generated by the permutation $\alpha = (1\ 2\ 3\ 4\ 5)$. Then clearly the permutation group (Γ, Y) is transitive because for any $i, j \in Y$ there is some power of α which maps i into j. But there is no power of α which, say, simultaneously maps 1 into 5 and 2 into 3. That is, not every ordered pair of distinct elements of Y can be mapped by a permutation in Γ into any other given ordered pair of distinct elements. We therefore say that the permutation group (G, Y) is not 2-transitive.

More generally, a permutation group (Γ, Y) is said to be k-*transitive* if, given any two k-tuples $(x_1, x_2, \ldots, x_k)$ and $(y_1, y_2, \ldots, y_k)$ of distinct elements of Y, then there is an $\alpha \in \Gamma$ such that

$$(\alpha(x_1), \alpha(x_2), \ldots, \alpha(x_k)) = (y_1, y_2, \ldots, y_k).$$

Thus, a transitive permutation group is 1-transitive. Also, (Γ, Y) is said to be k-*homogeneous* if, for any two k-subsets A, B of Y, there is an $\alpha \in \Gamma$ such that $\alpha(A) = B$, where $\alpha(A) = \{\alpha(a) : a \in A\}$.

Finally, let (Γ, Y) be transitive and suppose that $\mathcal{R}$ is an equivalence relation on Y, and let the equivalence classes of Y under $\mathcal{R}$ be $Y_1, Y_2, \ldots, Y_r$. Then (Γ, Y) is said to be *compatible* with $\mathcal{R}$ if, for any $\alpha \in \Gamma$ and any equivalence class Y_i, the set $\alpha(Y_i)$ is also an equivalence class. For example, if $Y = \{1, 2, 3, 4\}$ and Γ is the group generated by the permutation $(1\ 2\ 3\ 4)$, then (Γ, Y) is compatible with the relation whose equivalence classes are $\{1, 3\}$ and $\{2, 4\}$.

Any permutation group is clearly compatible with the trivial equivalence relations on Y, namely, those in which either all of Y is an equivalence class or when each singleton set is an equivalence class. If these are the only equivalence relations with which (Γ, Y) is compatible, then the permutation group is said to be *primitive*. Otherwise it is *imprimitive*.

If (Γ, Y) is imprimitive and $\mathcal{R}$ is a nontrivial equivalence relation on Y with which the permutation group is compatible, then the equivalence classes of $\mathcal{R}$ are called *imprimitivity blocks* and their set Y/R is an *imprimitivity block system* for the permutation group (Γ, Y).

It is an easy exercise (see Exercise 1.14) to show that a 2-transitive permutation group is primitive.

We shall also need some elementary ideas on the presentation of a group in terms of generators and relations.

Let Γ be a group and let $X \subseteq \Gamma$. A *word* in X is a product of a finite number of terms, each of which is an element of X or an inverse of an element of X. The set X is said to *generate* Γ if every element in Γ can be written as a word in X; in this case the elements of X are said to be *generators* of Γ. A *relation* in X is an equality between two words in X. By taking inverses, any relation can be written in the form $w = 1$, where w is some word in X.

If X generates Γ and every relation in Γ can be deduced from one of the relations $w_1 = 1, w_2 = 1, \ldots$ in X, then we write

$$\Gamma = \langle X | w_1 = 1, w_2 = 1, \ldots \rangle.$$

This is called a *presentation* of Γ in terms of generators and relations. The group Γ is said to be *finitely generated* (respectively, *finitely related*) if $|X|$ (respectively, the number of relations) is finite; it is called *finitely presented*, or we say that it has a *finite presentation*, if it is both finitely generated and finitely related.

It is clear that every finite group has a finite presentation (although the converse is false). Simply take $X = \Gamma$ and, as relations, take all expressions of the form $\alpha_i \alpha_j = \alpha_k$ for all $\alpha_i, \alpha_j \in \Gamma$. In other words, the multiplication table of Γ serves as the defining relations.

It is well to point out that removing relations from a presentation of a group in general gives a larger group, the extreme case being that of the *free group* which has only generators and no relations.

The simplest free group is the infinite cyclic group that has the presentation

$$\langle \alpha \rangle$$

with just one generator and no defining relation, whereas the cyclic group of order n has the presentation

$$\langle \alpha | \alpha^n = 1 \rangle;$$

this group is denoted by $\mathbb{Z}_n$.

The group with presentation

$$\langle \alpha, \beta \rangle$$

is the infinite free group on two elements. The dihedral group of degree n is denoted by D_n. It has order $2n$ and also has a presentation with two generators:

$$\langle \alpha, \beta | \alpha^2 = 1, \beta^n = 1, \alpha^{-1}\beta\alpha = \beta^{-1}\rangle.$$

Determining a group from a given presentation is not an easy problem. The reader who doubts this can try to show that the presentations

$$\langle \alpha, \beta : \alpha\beta^2 = \beta^3\alpha, \beta\alpha^2 = \alpha^3\beta\rangle$$

and

$$\langle \alpha, \beta, \gamma : \alpha^3 = \beta^3 = \gamma^3 = 1, \alpha\gamma = \gamma\alpha^{-1}, \alpha\beta\alpha^{-1} = \beta\gamma\beta^{-1}\rangle$$

both give the trivial group. We shall of course make a very simple use of standard group presentations where these difficulties do not arise. The book [159] is a standard reference for advanced work on group presentations.

The reader is referred to [147, 222] for any terms and concepts on group theory that are used but not defined in these chapters and, in particular, to [49, 62] for more information on permutation groups.

1.3 Graphs and groups

Let G, G' be two graphs. A bijection $\alpha : V(G) \to V(G')$ is called an *isomorphism* if

$$\{u, v\} \in E(G) \Leftrightarrow \{\alpha(u), \alpha(v)\} \in E(G').$$

The graphs G, G' are, in this case, said to be *isomorphic*, and this is denoted by $G \simeq G'$. Similarly, if D, D' are digraphs, then a bijection $\alpha : V(D) \to V(D')$ is called an *isomorphism* if

$$(u, v) \in A(D) \Leftrightarrow (\alpha(u), \alpha(v)) \in A(D'),$$

and in this case the digraphs D, D' are also said to be *isomorphic*, and again this is denoted by $D \simeq D'$.

If the two graphs, or digraphs, in this definition are the same, then α is said to be an *automorphism* of G or of D. The set of automorphisms of a graph or a digraph is a group under composition of functions, and it is denoted by Aut(G) or Aut(D).

Note that an automorphism α of G is an element of $S_{V(G)}$, although it is its induced action on $E(G)$ that determines whether α is an automorphism. This fact, although clear from the definition of automorphism, is worth emphasising when beginning to study automorphisms of graphs.

Figure 1.2. No automorphism permutes the edges as (12   23   34)

For example, for the graph in Figure 1.2, the permutation of edges given by (12   23   34) is not induced by any permutation of the vertex-set $\{1, 2, 3, 4\}$. The only automorphisms for this graph are the identity and the permutation (14)(23), which induces the permutation (12 34)(23) of the edges in the graph.

The question of edge permutations not induced by vertex permutations will be considered in some more detail later in this chapter.

The process of obtaining a permutation group from a digraph can be reversed in a very natural manner. Suppose that (Γ, Y) is a group of permutations acting on a set Y. Let A be a union of orbits of $(\Gamma, Y \times Y)$. Clearly, the digraph D whose vertex-set is Y and whose arc-set is A has Γ as a subgroup of its automorphism group. It might, however, happen that $\mathrm{Aut}(G)$ is larger than Γ. Moreover, if the pairs in A are such that, for every $(u, v) \in A$, (v, u) is also in A, then replacing every opposite pair of arcs of D by a single edge gives a graph G such that $\Gamma \subseteq \mathrm{Aut}(G)$.

This and other ways of constructing graphs or digraphs admitting a given group of permutations will be studied in more detail in Chapter 4.

Certain facts about automorphisms of graphs and digraphs are very easy to prove and are therefore left as exercises:

(i) $\mathrm{Aut}(G) = \mathrm{Aut}(\overline{G})$;
(ii) $\mathrm{Aut}(G) = S_{V(G)}$ if and only if G or $\overline{G}$ is K_n, the complete graph on n vertices;
(iii) $\mathrm{Aut}(C_n) = D_n$.

Also, let α be an automorphism of G and u, v vertices of G. Then,

(iv) $\deg(u) = \deg(\alpha(u))$;
(v) $G - u \simeq G - \alpha(u)$;
(vi) $d(u, v) = d(\alpha(u), \alpha(v))$, where $d(u, v)$ is the distance between u and v.

Also, if u is a vertex in a digraph D and α is an automorphism of D, then

(vii) $\deg_{\mathrm{in}}(u) = \deg_{\mathrm{in}}(\alpha(u))$ and $\deg_{\mathrm{out}}(u) = \deg_{\mathrm{out}}(\alpha(u))$.

If u and v are vertices in a graph G and there is an automorphism α of G such that $\alpha(u) = v$, then u and v are said to be *similar*. If $G - u \simeq G - v$, then u and v are said to be *removal-similar*. Property (v) tells us that if two vertices are similar, then they are removal-similar. The converse of this is, however,

false, as can be seen from the graph shown in Figure 1.3. Here, the vertices
u, v are removal-similar but not similar. Such vertices are called *pseudosimi-*
lar. Similar, removal-similar and pseudosimilar edges are analogously defined:
Two edges ab, cd of G are similar if there is an automorphism α of G such
that $\alpha(a)\alpha(b) = cd$. We shall be studying pseudosimilarity in more detail in
Chapter 5.

Sometimes we ask questions of this type: how many graphs (possibly of
some fixed order n) are there? The answer to this question depends heavily on
how we consider two graphs to be different.

In general, if the order of a graph G is n, we can think of its vertices as
being labelled with the integers $\{1, 2, \ldots, n\}$. Two graphs G and H of order n
so labelled are called *identical* or *equal as labelled graphs* (written $G = H$) if

$$ij \in E(G) \Leftrightarrow ij \in E(H).$$

(Compare this definition with that of isomorphic graphs.) Obviously, identical
graphs are isomorphic, but the converse is not true. For example, the graphs in
Figure 1.4 are isomorphic but not identical.

Counting nonisomorphic graphs is, in general, much more difficult than
counting nonidentical graphs. For example, there are four nonisomorphic graphs
on three vertices but eight nonidentical ones. These are shown in Figures 1.5
and 1.6, respectively.

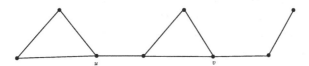

Figure 1.3. A pair of pseudosimilar vertices

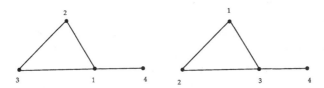

Figure 1.4. Isomorphic but nonidentical graphs

Figure 1.5. The four nonisomorphic graphs of order 3

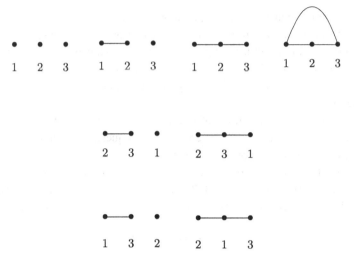

Figure 1.6. The eight nonidentical graphs of order 3

Counting nonisomorphic graphs involves consideration of group symmetries. For more on this the reader is referred to [103].

1.4 Edge-automorphisms and line-graphs

Although we shall be dealing mostly with Aut(G) and its realisation as the permutation group (Aut(G), $V(G)$), let us briefly look at other related groups associated with G. In this section we shall assume that G is a nontrivial graph, that is, its edge-set is nonempty.

An *edge-automorphism* of a graph G is a bijection θ on $E(G)$ such that two edges e, f are adjacent in G if and only if $\theta(e), \theta(f)$ are also adjacent in G. The set of all edge-automorphisms of G is a group under composition of functions, and it is denoted by $\text{Aut}_1(G)$.

The concept of edge-automorphisms can perhaps be best understood within the context of line-graphs. The *line-graph* $L(G)$ of a graph G is defined as the graph whose vertex-set is $E(G)$ and in which two vertices are adjacent if and only if the corresponding edges are adjacent in G. An automorphism of $L(G)$ is clearly an edge-automorphism of G and (Aut$_1(G), E(G)$) is equivalent to (Aut($L(G)$), $V(L(G))$). In this section we shall give the exact relationship between Aut$_1(G)$ and Aut(G), that is, between the automorphism groups of G and $L(G)$.

As we described earlier, any automorphism α of G naturally induces a bijection $\hat{\alpha}$ on $E(G)$ defined by $\hat{\alpha}(uv) = \alpha(u)\alpha(v)$. It is an important (and easy to verify) property of $\hat{\alpha}$ that two edges e_1, e_2 are adjacent if and only if $\hat{\alpha}(e_1), \hat{\alpha}(e_2)$ are adjacent, that is, if and only if $\hat{\alpha}$ is an edge-automorphism. For this reason $\hat{\alpha}$ is called an *induced edge-automorphism* of G.

The set of all induced edge-automorphisms of G is denoted by $\text{Aut}^*(G)$, and it is easy to check that this is a subgroup of $\text{Aut}_1(G)$ under composition of functions. Now, it seems natural to expect that $\text{Aut}(G)$ and $\text{Aut}^*(G)$ are isomorphic. However, it can happen that two different automorphisms of G induce the same edge-automorphism. For example, let $G = K_2$. Then $|\text{Aut}(G)| = 2$ but $|\text{Aut}^*(G)| = 1$. Also, suppose that G contains isolated vertices. Then any automorphism of G that permutes the isolated vertices and leaves all of the others fixed induces the trivial edge-automorphism. The following theorem says that these are basically the only situations when $\text{Aut}(G) \not\simeq \text{Aut}^*(G)$.

Theorem 1.2 *Let G be a nontrivial graph. Then* $\text{Aut}(G) \simeq \text{Aut}^*(G)$ *if and only if G has at most one isolated vertex and K_2 is not a component.*

Proof Clearly, the mapping $\alpha \mapsto \hat{\alpha}$ is a homomorphism from $\text{Aut}(G)$ onto $\text{Aut}^*(G)$ because $\hat{\alpha}.\hat{\beta}(uv) = \alpha.\beta(u)\alpha.\beta(v) = \widehat{\alpha\beta}(uv)$. We must therefore show that the kernel of this mapping is trivial if and only if G has at most one isolated vertex and K_2 is not a component.

Suppose first that G has two isolated vertices u, v or K_2 as a component with vertices u, v. Then the permutation α that transposes u and v and fixes all of the other vertices is a nontrivial automorphism of G, but $\hat{\alpha}$ is the identity. Therefore the kernel is not trivial.

Conversely, suppose that G does not contain K_2 as a component nor its complement. If $\text{Aut}(G)$ is trivial, then so is $\text{Aut}^*(G)$. Therefore, let α be a nontrivial element of $\text{Aut}(G)$, and let $\alpha(u) = v \neq u$. Then $\deg(u) = \deg(v) \neq 0$ (otherwise u, v would be a pair of isolated vertices). We consider two cases.

Case 1: u, v adjacent. Let e be the edge uv. Then $\deg(u) = \deg(v) > 1$ (otherwise, the two vertices u, v would form a component K_2). Therefore, there exists an edge $f \neq e$ incident to u (but not to v, since the graph is simple). But $\hat{\alpha}(f)$ must be incident to v (since $\alpha(u) = v$), that is, $\hat{\alpha}(f) \neq f$, and hence $\hat{\alpha}$ is not trivial.

Case 2: u, v not adjacent. Let e be an edge incident to u. Again, e is not incident to v but $\hat{\alpha}(e)$ is. Therefore $\hat{\alpha}$ is again nontrivial. □

The next natural question to ask is whether there can be edge-automorphisms of G that are not induced by automorphisms, that is, whether $\text{Aut}^*(G)$ can be a strict subgroup of $\text{Aut}_1(G)$. This situation can very well happen, although, as we shall see, such cases are quite rare.

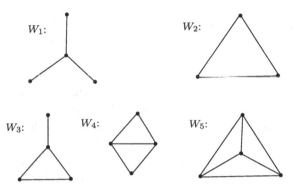

Figure 1.7. Graphs with edge-isomorphisms not induced by isomorphisms

Before proceeding let us first extend the idea of edge-automorphisms on the edge-set of a graph to that of edge-isomorphisms between edge-sets of different graphs.

Let G, G' be two nontrivial graphs. A bijection $\theta : E(G) \rightarrow E(G')$ is an *edge-isomorphism* if

$$e, f \text{ adjacent in } G \Leftrightarrow \theta(e), \theta(f) \text{ adjacent in } G'.$$

Two graphs are said to be *edge-isomorphic* if there is an edge-isomorphism between their edge-sets.

The graphs W_1, W_2 in Figure 1.7 are edge-isomorphic, although they are not isomorphic. That is, there is an edge-isomorphism between their edge-sets that cannot be induced by an isomorphism between their vertex-sets. This means that their line-graphs are isomorphic even though the two graphs are not themselves isomorphic.

Also, each of the graphs W_3, W_4, W_5 in the same figure has edge-automorphisms that are not induced by automorphisms. That is, the group $\text{Aut}^*(W_i)$ is a strict subgroup of $\text{Aut}_1(W_i)$. In other words, $\text{Aut}(L(W_i))$ is larger than $\text{Aut}(W_i)$.

The following theorem of Whitney [258] says that these are essentially the only cases when edge-isomorphisms that are not induced by isomorphisms can arise. We give the statement of the theorem without proof, which, although not deep or difficult, would lengthen this introductory chapter without adding significant new insights.

Theorem 1.3 (Whitney) *Let G, G' be connected graphs different from the five graphs in Figure 1.7. Let $\theta : E(G) \rightarrow E(G')$ be an edge-isomorphism. Then θ is induced by an isomorphism from G to G'.*

Whitney's Theorem and Theorem 1.2 together therefore give the following corollary.

Corollary 1.4 *Let G be a nontrivial graph. Then* $\mathrm{Aut}_1(G) = \mathrm{Aut}^*(G)$ *if and only if both of these conditions hold:*

    (i) *not both W_1, W_2 are components of G;*

    (ii) *none of $W_i, i = 3, 4, 5$ is components of G.*

Moreover, $\mathrm{Aut}_1(G) \simeq \mathrm{Aut}(G)$ *(that is, $\mathrm{Aut}(L(G)) \simeq \mathrm{Aut}(G)$) if and only if* (i) *and* (ii) *hold and G has at most one isolated vertex and K_2 is not a component of G.*

1.5 A word on issues of computational complexity

Although in this book we shall not concern ourselves with issues of computational complexity, it is perhaps worthwhile to say a few words in this regard here in order to put matters into a better perspective. A student reading the definitions of isomorphic graphs and automorphisms might think that it is an easy matter to determine in general whether two given graphs are isomorphic or to compute the automorphism group of a graph. In fact, this is far from being the case, and these problems are very hard to crack in practice, at least as far as present knowledge goes.

In general, one considers that an efficient algorithm exists for finding a solution to a problem (for example, finding a nontrivial automorphism of a given graph) if there is a general algorithm such that the number of operations that it takes to solve the problem is a polynomial function of the size of the input (say, the number of vertices in the graph); one says that the algorithm solves the problem in polynomial time.

Of course, several terms in the previous sentence need exact definitions, but we shall here take an intuitive approach and refer the reader to [36] or [82] for the exact details on computational complexity.

Those problems for which an efficient (polynomial-time) algorithm exists form the class denoted by P (which stands for 'polynomial'). However, there are several problems for which it is not known whether an efficient algorithm does exist. In order to tackle this question of computational intractibility, two important ideas have been developed.

Firstly, the class NP (which stands for 'nondeterministic polynomial') is defined. Roughly (again we refer the reader to the textbooks cited earlier for the exact details) this class contains all of those problems for which, given a candidate solution, one can verify in polynomial time that it is in fact a correct solution. For example, the problem of determining whether a graph has a non-trivial automorphism is in NP, since, given such a permutation of the vertices, it is easy to determine in polynomial time that it is an automorphism.

Now the main question in computational complexity is whether P = NP (clearly P ⊆ NP), and to tackle this question another important idea is introduced. Given two problems A and B, one says that A is (polynomially) reducible to B if, given an algorithm for solving B, it can be transformed in polynomial time into an algorithm for solving A. Reducibility therefore introduces a hierarchy between problems for, if A is reducible to B, then, in a sense, A cannot be more difficult to solve (computationally) than B. In particular, if there is an efficient algorithm for solving B, then there is also an efficient algorithm for solving A.

Now, the question of reducibility took on special significance by the discovery that in the class NP there are problems, called NP-complete, to which any other problem in NP is reducible. In other words, if an efficient algorithm can be found for any NP-complete problem, then all problems in NP would have an efficient algorithm to solve them, and P would be equal to NP.

Now, it is not known whether the problem of determining if two graphs are isomorphic, which lies clearly in NP, is NP-complete. In fact, if it turns out that P is not equal to NP, then there is evidence to suggest that the problem of graph isomorphism might lie strictly between the classes P and NP.

What all this means in practice is that, as far as present knowledge goes, no general algorithm can determine in a guaranteed reasonable time whether two graphs are isomorphic, or whether a given graph has a nontrivial automorphism (these two problems are closely related [126]). It is known that for special types of graphs (for example, trees, planar graphs and graphs with bounded degree) an efficient algorithm does exist.

Computer packages can also help one to solve these problems, certainly more efficiently than an attempt 'by hand' for large graphs, although their time performance is not guaranteed (by what we said earlier). For example, the software package *Mathematica*^{TM} has a combinatorics extension that, amongst other things, finds graph automorphisms and isomorphisms. A more specialised package, and one that is freely available from

www.combinatorialmath.org.ca/g&g/index.html

is *Groups & Graphs* [131], developed by Bill Kocay. This package contains several combinatorial routines related to graphs, digraphs, combinatorial designs and their automorphism groups and also embeddings of graphs on some surfaces and a graph isomorphism algorithm. It is easy to use, and it has a pleasant graphical user interface. It is also very useful simply for drawing diagrams of graphs. Although originally written for Macintosh^{TM} computers, a version for the unix-based Haiku operating system is in preparation, and this version will contain several new features.

An important computer algebra package, which is also freely available, is the system GAP [243]. This package performs very sophisticated routines in discrete abstract algebra, in particular routines on permutation groups. It incorporates a number of extensions, one of which, GRAPE [235], deals specifically with graphs, including their automorphisms and isomorphisms.

The computer package *Sage* [227] is an open-source competitor to systems like Maple^{TM}, Mathematica^{TM} and Matlab^{TM}. It incorporates several open-source mathematical software like GAP and R, and it can be run via Sage-MathCloud without the need of installing the system on one's computer. It has an excellent library of functions for doing graph theory. In this book we shall present some constructions using GAP and *Sage*.

Finally, it should be mentioned that it is generally accepted that the best package to tackle graph isomorphisms is *nauty* [181], developed by Brendan McKay. In fact, the system GRAPE invokes *nauty* when computing automorphisms or isomorphisms.

1.6 Exercises

1.1 Draw all twenty nonidentical graphs with vertex-set $\{1, 2, 3, 4\}$ that have three edges. How many of them are nonisomorphic? In general, how many nonidentical graphs on n vertices and m edges are there? How many are there on n vertices?

1.2 Let G be the graph in Figure 1.8. How many nonidentical labellings does G have using the labels $\{1, 2, \ldots, 6\}$ on its vertices?

In general, how many nonidentical labellings does a graph G on n vertices have using the labels $\{1, 2, \ldots, n\}$ on its vertices?

1.3 Show that if G is *self-complementary* (that is, $G \simeq \overline{G}$), then $n \equiv 0 \bmod 4$ or $n \equiv 1 \bmod 4$. Determine all self-complementary graphs on five vertices.

1.4 A well-known result due to Cayley says that the number of nonidentical trees on n vertices is n^{n-2}. Verify this for $n = 4$. Look up one of the several proofs of this result.

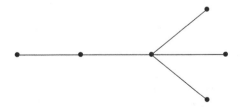

Figure 1.8. How many distinct labellings does this graph have?

The points $1, 2, \ldots, n$ are drawn in a plane. A random tree is drawn joining these points, with all possible spanning trees being equally likely. Let p_n be the probability that 1 is an endvertex of the tree. Show that $\lim_{n \to \infty} p_n = 1/e$.

1.5 Find a graph with a pair of pseudosimilar edges.

1.6 Let $Y = \{1, 2, 3, 4\}$ and let Γ be the group acting on Y generated by the permutation $(1\ 2\ 3\ 4)$. Construct a digraph D whose vertex-set is Y and whose arc-set is the orbit of the arc $(1, 2)$ under the action of the permutation group $(\Gamma, Y \times Y)$. Is Γ the whole of $\text{Aut}(D)$? Can a graph be obtained by taking the orbit of some other arc or a union of orbits? Will Γ always be the whole of $\text{Aut}(D)$?

1.7 Let P be a rectangular plate and Q a plate in the form of a rhombus. Show that the groups of symmetry of P and Q are isomorphic as abstract groups but not equivalent considered as permutation groups of the four vertices of P and Q.

Show that the same situation arises with the following two graphs: the cycle on four vertices with an extra multiple edge and the complete graph K_4 with an edge deleted.

1.8 Show that the dihedral group D_n can be presented as

$$\langle \alpha, \beta \,|\, \alpha^2 = \beta^2 = (\alpha\beta)^n = 1 \rangle.$$

1.9 Let Γ_1 be the abelian group defined by the presentation

$$\langle \alpha, \beta, \gamma \,|\, \alpha^3 = \beta^3 = \gamma^3 = 1, [\alpha, \beta] = [\alpha, \gamma] = [\beta, \gamma] = 1 \rangle,$$

where $[\alpha, \beta] = \alpha^{-1}\beta^{-1}\alpha\beta$ is the commutator of α and β, and let Γ_2 be the group defined by the presentation

$$\langle \alpha, \beta, \gamma \,|\, \alpha^3 = \beta^3 = \gamma^3 = 1, [\beta, \alpha] = \gamma, [\alpha, \gamma] = [\beta, \gamma] = 1 \rangle.$$

Show that although Γ_1 and Γ_2 are not isomorphic as abstract groups, and therefore the two permutation groups $(L(\Gamma_1), \Gamma_1)$ and $(L(\Gamma_2), \Gamma_2)$ are not equivalent, still the cycle structures of the permutations of $L(\Gamma_1)$ acting on Γ_1 are the same as those of the permutations of $L(\Gamma_2)$ acting on Γ_2.

Give an example of two permutation groups whose permutations have the same cycle structures and which are isomorphic as abstract groups but are still not equivalent as permutation groups.

1.10 Prove Theorem 1.1.

1.11 This exercise is intended to illustrate Theorem 1.1. Consider the action of the alternating group A_4 on the set $X = \{1, 2, 3, 4\}$. Let $\mathcal{H}$ be the stabiliser of the element 1 under this action. Confirm that the left action of A_4 on the left cosets of $\mathcal{H}$ is equivalent to the action of A_4 on X.

1.12 Let Γ be a group of permutations acting on a set Y and let y, x be two elements of Y that are in the same orbit under this action. Let $\alpha, \beta \in \Gamma$ be two permutations such that $\alpha(y) = \beta(y) = x$. Prove that $\alpha\Gamma_y = \beta\Gamma_y$.

Prove also that if x' is in the same orbit as y and γ is a permutation such that $\gamma(y) = x'$ and $\gamma\Gamma_y = \alpha\Gamma_y$, then $x' = x$.

1.13 Show that if the abelian permutation group Γ acts transitively on Y, then its action on Y is regular.

1.14 (a) Show that a 2-transitive permutation group is primitive.
(b) Show that if $\text{Aut}(G)$, for a graph G, is 2-transitive, then either G or its complement is a complete graph.

(c) Suppose that the transitive permutation group (Γ, Y), with $|Y|$ finite, is imprimitive. Show that the blocks of an imprimitivity block system have equal size.

1.15 Let G be a vertex-transitive graph whose automorphism group acts imprimitively on $V(G)$. Show that the subgraphs of G induced by the blocks of an imprimitivity block system are all isomorphic.

   Suppose that each such subgraph is replaced by its complement, leaving the other edges intact. Let G' be the resulting graph. Show that $\text{Aut}(G') = \text{Aut}(G)$.

1.16 The Petersen graph can be defined as follows. Let $N = \{1, 2, 3, 4, 5\}$, and let the vertices of the graph be all subsets of N of size 2 in which two vertices are adjacent if the corresponding subsets are disjoint. Use this definition and GAP (with GRAPE) to construct the Petersen graph and to verify some of its properties.

1.7 Notes and guide to references

One of the standard texts on graph theory has, for many years, been [97]. More recent books that give an excellent coverage of the subject are [28, 61, 257, 259]. The last reference is a short introduction that is quite sufficient background for this book. Biggs' book [24] is the standard text on algebraic graph theory, but the more recent [90] is also an excellent and up-to-date textbook on the subject. The book [94] contains a number of recent and specialised survey papers on various aspects of algebraic graph theory, particularly those dealing with graph symmetries. A proof of Whitney's Theorem can be found in [22].

We shall need only the most elementary notions of group theory. The text [147] gives ample coverage for our purposes, while [222] provides a more complete treatment. Two excellent books devoted entirely to permutation groups are [49, 62]. Most of the results and definitions on permutation groups that we have given here and others that we shall need can be found in the first few chapters of these two books.

For a full discussion of the terms on computational complexity that were introduced earlier rather intuitively, the reader is referred to the standard textbook [82] or the more recent [36]. The book [126] and the references it cites are suggested for those who are interested in the computational complexity of the graph isomorphism problem. Those who are particularly interested in some of the powerful algebraic techniques used to tackle this problem should look at the papers [106, 155]. For practical computations on a computer with permutation groups and graph automorphisms and isomorphisms in particular, the systems [131, 181, 235, 243] are recommended.

2

Various Types of Graph Symmetry

We shall see in this chapter that most graphs are *asymmetric*, that is, their automorphism group is trivial; in other words, it consists only of the identity permutation. The least number of vertices that an asymmetric graph can have is six, and the graph shown in Figure 2.1 is the smallest such graph in the sense that any other asymmetric graph on six vertices has more edges.

As is to be expected, however, the most interesting relationships between groups and graphs arise when the graphs have a very high degree of symmetry, that is, a large automorphism group. One way to make more precise the idea of a large automorphism group is to require that it at least be transitive on the vertex-set or the edge-set of the graph.

The weakest forms of symmetry to ask of a graph involve vertex-transitivity and edge-transitivity, which we define in the next section.

2.1 Transitivity

Recall the definition of similar vertices from the previous chapter. We say that a graph G is *vertex-transitive* if any two vertices of G are similar, that is, if, for any $u, v \in V(G)$, there is an automorphism α of G such that $\alpha(u) = v$. In other words, G is vertex-transitive if all of the vertices of G are in the same orbit of the permutation group $(\mathrm{Aut}(G), V(G))$.

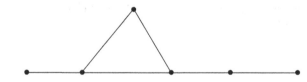

Figure 2.1. The smallest asymmetric graph

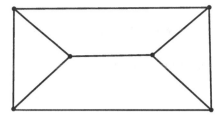

Figure 2.2. A vertex-transitive graph that is not edge-transitive

One can define edge-transitivity analogously. A graph G is *edge-transitive* if, given any two edges $\{a, b\}$ and $\{c, d\}$, there exists an automorphism α such that $\alpha\{a, b\} = \{c, d\}$, that is, $\{\alpha(a), \alpha(b)\} = \{c, d\}$. In other words, G is edge-transitive if any two of its edges are similar under the action of the permutation group $(\text{Aut}^*(G), E(G))$, that is, if the edges of G are all in one orbit under this action.

Note that very often the word 'transitive' is used to refer to a graph, and in this case it is taken to mean 'vertex-transitive'.

Vertex-transitivity does not imply edge-transitivity, nor does the converse implication hold. Figure 2.2 shows a graph that is vertex-transitive but not edge-transitive.

The complete bipartite graph $K_{p,q}$ with $p \neq q$ is a simple example of a graph that is edge-transitive but not vertex-transitive. The following well-known result does give a description of edge-transitive graphs that are not vertex-transitive.

Theorem 2.1 *Let G be a graph without isolated vertices and let $\mathcal{H}$ be a subgroup of $\text{Aut}(G)$. Suppose that the action induced by $\mathcal{H}$ is transitive on the edges of G but not on its vertices. Then G is bipartite and the action of $\mathcal{H}$ on $V(G)$ has two orbits that form the bipartition of $V(G)$.*

Proof Let $\{u, v\}$ be an edge of G. Let V_1, V_2 be the orbits under the action of $\mathcal{H}$ containing u and v, respectively. (We are not excluding, for the moment, the possibility that $V_1 = V_2$.) Let x be any other vertex of G. Since G does not have isolated vertices, there exists a vertex y adjacent to x; that is, $\{x, y\}$ is an edge of G. But G is edge-transitive under the action of $\mathcal{H}$; therefore the two edges are similar under this action. Hence x is similar to at least one of u or v, that is, x is in V_1 or V_2. Therefore $V_1 \cup V_2 = V(G)$.

Now, V_1, V_2 must be disjoint; otherwise (since orbits form a partition) they are equal, and this would mean that the vertices of G are all in one orbit, giving that G is vertex-transitive under the action of $\mathcal{H}$. Hence we now have that the action of $\mathcal{H}$ on $V(G)$ has exactly two orbits, V_1, V_2.

Now let a, b be in the same orbit, say V_1. It then follows that a, b are not adjacent. For suppose otherwise. Then the edge $\{u, v\}$ is similar to the edge $\{a, b\}$ under the action of $\mathcal{H}$; therefore the vertex v is similar to one of a or b under this action, giving that v (which is in the orbit V_2) is also in the orbit V_1, which contains a and b. But this is impossible since $V_1 \cap V_2 = \emptyset$.

Hence, as required, we have that no two vertices from the same orbit can be adjacent. $\square$

Corollary 2.2 *If a graph G without isolated vertices is edge-transitive but not vertex-transitive, then it is bipartite and the action of* Aut(G) *on* V(G) *has two orbits that form the bipartition of* V(G).

Proof Take $\mathcal{H} = \text{Aut}(G)$ in the previous theorem. $\square$

2.1.1 Semisymmetric graphs

The typical example of edge-transitive but not vertex-transitive graphs given earlier is the complete bipartite graph with a different number of vertices in the bipartition. These graphs are trivially not vertex-transitive because their vertices have different degrees. Although regular graphs do exist that are edge-transitive but not vertex-transitive, it is quite difficult to construct them. Such graphs are now called *semisymmetric graphs* and they were first studied by Folkman [77], who constructed the smallest possible semisymmetric graph having twenty vertices.[1] One construction of the Folkman Graph is described in Exercise 2.5.

Here we shall describe another well-known semisymmetric graph, the Gray Graph.[2] For a long time nobody could find a smaller cubic semisymmetric graph, and eventually it was formally proved in [160] that it is *the* smallest cubic semisymmetric graph. We shall see a more systematic way of describing it in subsequent chapters. Here we follow Bouwer's construction in [33].

Consider a cycle on 54 vertices numbered consecutively from 0 to 53. To form the Gray Graph G add the following edges to this cycle:

$$\{1, 42\}, \{2, 15\}, \{3, 28\}, \{4, 33\}, \{5, 44\}, \{6, 53\}, \{7, 48\}, \{8, 21\}, \{9, 32\},$$

$$\{10, 45\}, \{11, 24\}, \{12, 41\}, \{13, 20\}, \{14, 31\}, \{16, 35\}, \{17, 40\}, \{18, 49\},$$

$$\{19, 0\}, \{22, 51\}, \{23, 30\}, \{25, 38\}, \{26, 43\}, \{27, 52\}, \{29, 36\}, \{34, 47\},$$

---

[1] More information about this graph, called the Folkman Graph, can be found on the MathWorld page http://mathworld.wolfram.com/FolkmanGraph.html

[2] The Gray Graph is also featured on MathWorld at http://mathworld.wolfram.com/Gray Graph.html

$$\{37, 50\}, \{39, 46\}.$$

It is tedious but not difficult to check that the permutations

$$\alpha = (2\ 0\ 43)(3\ 53\ 43)(4\ 6\ 44)(7\ 45\ 33)(8\ 10\ 32)(11\ 31\ 21)$$
$$= (12\ 14\ 20)(15\ 19\ 41)(16\ 18\ 40)(22\ 24\ 30)(25\ 29\ 51)$$
$$= (26\ 28\ 52)(34\ 48\ 46)(35\ 49\ 39)(36\ 50\ 38)$$

and

$$\beta = (1\ 7\ 11\ 37\ 15\ 53\ 9\ 25\ 35)(2\ 6\ 10\ 38\ 16\ 0\ 8\ 24\ 36)$$
$$= (3\ 5\ 45\ 39\ 17\ 19\ 24\ 23\ 29)(4\ 44\ 46\ 40\ 18\ 20\ 22\ 30\ 28)$$
$$= (12\ 50\ 14\ 52\ 32\ 26\ 34\ 42\ 48)(13\ 51\ 31\ 27\ 33\ 43\ 47\ 41\ 49)$$

are automorphisms of G.

Note that the automorphism α fixes the vertex 1 and permutes cyclically its neighbours 2, 42 and 0. Thus, in order to show that the graph is edge-transitive it is sufficient to show that any odd-numbered vertex can be mapped into 1 by an automorphism of G. This can be done by appropriate products of α and β. For example, $\alpha^4\beta$ maps vertex 53 to vertex 1.

However, the graph is not vertex-transitive because from an odd-numbered vertex it is possible to have three different paths of length 4 joining the vertex to some other common vertex (for example, vertex 1 to vertex 5), but this is not possible from an even-numbered vertex.

Another way to show that the Gray Graph is not vertex-transitive is to consider the distance sequences of its vertices [166]. The distance sequence of a vertex v is the vector $(a_0, a_1, \ldots, a_r)$ where a_i is the number of vertices at distance i from v. In the case of the Gray Graph, the distance sequences of the vertices in the two colour classes are $(1, 3, 6, 12, 12, 12, 8)$ and $(1, 3, 6, 12, 16, 12, 4)$, respectively, therefore these vertices cannot be in the same orbit under the automorphism group of the graph.

Although the *Sage* package has the Gray Graph already implemented, it is easy and instructive to show how to construct it following the specification given earlier. First one creates the vertex-set which will be the list of numbers from 1 to 54. *Sage*, like many computer languages such as Python, starts its lists from 0. Therefore a list of length n produced by the command `range(55)` would contain the numbers from 0 to 54. To start the list from 1 we have to define the list of vertices as

```
vertices := range(1,55);
```

The graph is then constructed first using the command DiGraph so that
we do not need to repeat every pair of adjacent vertices twice. This commands basically takes two parameters. The first parameter is the vertex-set
of the graph to be constructed, and the second parameter is a boolean function
(defined with the lambda construct) of two variables which compares all possible pairs of the vertex-set and an edge is drawn between any pair of vertices
for which the function returns True.

```
dgray := DiGraph([vertices, lambda i, j:
                  (i == mod[j + 1, 54]) or
                  (i == 53 and j == 54)  or
                  (i == 1 and j == 42)   or
                  (i == 2 and j == 15)   or
                  (i == 3 and j == 28)   or
                  (i == 4 and j == 33)   or
                  (i == 5 and j == 44)   or
                  (i == 6 and j == 53)   or
                  (i == 7 and j == 48)   or
                  (i == 8 and j == 21)   or
                  (i == 9 and j == 32)   or
                  (i == 10 and j == 45)  or
                  (i == 11 and j == 24)  or
                  (i == 12 and j == 41)  or
                  (i == 13 and j == 20)  or
                  (i == 14 and j == 31)  or
                  (i == 16 and j == 35)  or
                  (i == 17 and j == 40)  or
                  (i == 18 and j == 49)  or
                  (i == 19 and j == 54   or
                  (i == 22 and j == 51)  or
                  (i == 23 and j == 30)  or
                  (i == 25 and j == 38)  or
                  (i == 26 and j == 43)  or
                  (i == 27 and j == 52)  or
                  (i == 29 and j == 36)  or
                  (i == 34 and j == 47)  or
                  (i == 37 and j == 50)  or
                  (i == 39 and j == 46) ] )
```

This digraph is then changed into an undirected graph with the following
command which changes every arc into an edge.

```
gray = dgray.to_undirected()
```

It is then easy to check that the aforementioned properties of the Gray Graph hold. For example, in order to check whether it is vertex-transitive we use the command

```
gray.is_vertex_transitive()
```

which returns False. The command

```
gray.is_edge_transitive()
```

returns True, as expected, while the command

```
gray.is_regular()
```

also returns True, confirming that the graph is semisymmetric. In fact, we could have reached the same conclusion with the command

```
gray.is_semi_symmetric()
```

which also returns True.

Finally, one can check whether the graph constructed earlier is isomorphic to *Sage*'s inbuilt 'GrayGraph' using the command

```
graphs.GrayGraph.is_isomorphic(gray)
```

which, again, returns True.

We shall have more to say about the Gray Graph in a later chapter when we shall describe it in a more algebraic fashion.

Exercises 2.2 and 2.5 show that any semisymmetric graph must have even order and its degree must be less than $|V(G)|/2$.

2.1.2 Arc-transitive and $\frac{1}{2}$-arc-transitive graphs

A stronger form of transitivity than either vertex- or edge-transitivity based on the edge-set of G can also be defined. If G has the property that, for any two edges $\{a, b\}, \{c, d\}$, there is an automorphism α such that $\alpha(a) = c$ and $\alpha(b) = d$ and also an automorphism β such that $\beta(a) = d$ and $\beta(b) = c$, then G is said to be *arc-transitive*.

We shall now derive a result of Tutte that gives a restriction on the degree of the vertices of a graph that is vertex-transitive and edge-transitive but not arc-transitive.

One can think of arc-transitivity as follows. Given any graph G, construct the directed graph $\overleftrightarrow{G}$ obtained from G by replacing each edge $\{a, b\}$ by the pair of arcs (a, b) and (b, a). Then clearly $\mathrm{Aut}(G) = \mathrm{Aut}(\overset{\leftrightarrow}{G})$ and any automorphism α of G induces the natural action on arcs given by

$$(a, b) \mapsto (\alpha(a), \alpha(b)).$$

Then G is arc-transitive precisely if, given any two arcs in $\overset{\leftrightarrow}{G}$, there is an automorphism of $\overset{\leftrightarrow}{G}$ mapping one arc into the other.

This is a stronger form of transitivity than both vertex- and edge-transitivity because now, given any two edges on each of which an orientation is imposed, there is an automorphism mapping one edge into the other and preserving the given orientations. In fact, an arc-transitive graph is both vertex-transitive and edge-transitive.

Lemma 2.3 *Let $\mathcal{H}$ be a subgroup of $\mathrm{Aut}(G)$ such that, under the action of $\mathcal{H}$, G is vertex-transitive and edge-transitive but not arc-transitive. Let t be an arc of $\overset{\leftrightarrow}{G}$ and let D be the subdigraph of $\overset{\leftrightarrow}{G}$ whose vertex-set is $V(\overset{\leftrightarrow}{G})$ and whose arc-set is the orbit of t under the action of $\mathcal{H}$. Then*

(i) *for every edge $\{a, b\}$ of G, D contains exactly one of the arcs (a, b) or (b, a);*

(ii) *$\mathcal{H} \leq \mathrm{Aut}(D)$;*

(iii) *D is vertex-transitive.*

Proof (i) Let $t = (s_1, s_2)$. Because G is edge-transitive under the action of $\mathcal{H}$, there is an $\alpha \in \mathcal{H}$ such that $\alpha\{s_1, s_2\} = \{a, b\}$. Therefore certainly one of (a, b) or (b, a) is an arc of D. Suppose that both are arcs of D. Then there is some $\beta \in \mathcal{H}$ such that $\beta((a, b)) = (b, a)$. But given any edge $\{c, d\}$ of G there is, by edge-transitivity, a $\gamma \in \mathcal{H}$ such that $\gamma((a, b))$ equals (c, d) or (d, c). Suppose, without loss of generality, that $\gamma((a, b)) = (c, d)$. But then $\gamma\beta((a, b)) = (d, c)$. Therefore, for any edge $\{c, d\}$ of G, both arcs (c, d) and (d, c) are in the same orbit, that is, the action of $\mathcal{H}$ on G is arc-transitive, a contradiction.

(ii) This follows because the arc-set of D is a full orbit of the permutation group $(\mathcal{H}, V(G) \times V(G))$.

(iii) This follows because $(\mathcal{H}, V(G))$ is transitive, $V(D) = V(G)$ and $\mathcal{H} \leq \mathrm{Aut}(D)$. $\qquad\square$

If we let $\mathcal{H} = \mathrm{Aut}(G)$ in this lemma, then, in view of (i), if G is a vertex-transitive and edge-transitive graph that is not arc-transitive, it follows that the arc-set of $\overset{\leftrightarrow}{G}$ is naturally partitioned into two orbits of equal size under the action of $\mathrm{Aut}(G)$, and none of the two orbits contains both an arc (a, b) and its inverse (b, a). In view of this, a graph that is vertex-transitive, edge-transitive but not arc-transitive is said to be $\frac{1}{2}$-*arc-transitive*.

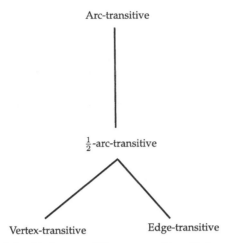

Figure 2.3. Relationship between different types of transitivity

The relationship between these forms of transitivity is shown in Figure 2.3, where a line leading down from one property to another means that the first implies the second.

Although $\frac{1}{2}$-arc-transitive graphs are not easy to find, they do exist (an example will be given in Chapter 3). The following well-known theorem of Tutte tells us that such a graph must have even degree.

Theorem 2.4 (Tutte) *Let $\mathcal{H}$ be a subgroup of* Aut(G) *such that, under the action of $\mathcal{H}$, G is vertex-transitive and edge-transitive but not arc-transitive. Then the degree of G is even. In particular, a $\frac{1}{2}$-arc-transitive graph has even degree.*

Proof   Let D be as in the previous lemma. By the third part of this lemma, all vertices of D have the same out-degree, say k. Now, $k \cdot |V(D)| = |A(D)|$ and, by the first part of the lemma, $|A(D)| = |E(G)|$. But if the common degree of the vertices of G is d, then, by the Handshaking Lemma, $|E(G)| = d \cdot |V(G)|/2 = d \cdot |V(D)|/2$. Therefore $d = 2k$, that is, d is even. $\square$

2.2 Asymmetric graphs

Although we shall be mostly interested in graphs with nontrivial automorphism groups, let us briefly consider asymmetric graphs. Let $\mathcal{P}$ be a graph theoretic property such as 'planar' or 'vertex-transitive'. Let r_n denote the proportion of labelled graphs on n vertices that have property $\mathcal{P}$. If $\lim_{n \to \infty} r_n = 1$, then we say that *almost every (a.e.) graph has property* $\mathcal{P}$.

We have already said that almost every graph is asymmetric. We shall soon prove a stronger result that will be used in a later chapter when we consider the Reconstruction Problem.

The following probability space is often set up when studying random graphs. Let $\mathcal{G}(n,p)$ be the set of all labelled graphs on the set of vertices $\{1, 2, \ldots, n\}$ where, for each pair i, j,

$$P(ij \text{ is an edge}) = p$$

and

$$P(ij \text{ is not an edge}) = 1 - p$$

independently. Therefore a graph with m edges in $\mathcal{G}(n,p)$ has probability $p^m q^{\binom{n}{2}-m}$, where $q = 1 - p$. We shall need only this space when the probability $p = \frac{1}{2}$. In this case, each graph G in $\mathcal{G}(n, \frac{1}{2})$ has probability $(\frac{1}{2})^{\binom{n}{2}}$, which is, of course, equal to the probability of choosing G randomly from amongst all $2^{\binom{n}{2}}$ labelled graphs on n vertices when all are equally likely to be chosen. Therefore, to show that a.e. graph has a particular property $\mathcal{P}$ one has to show that the probability that $G \in \mathcal{G}(n, \frac{1}{2})$ has property $\mathcal{P}$ tends to 1 as n tends to infinity.

Now, let k be fixed. We say that a graph G has *property A_k* if all induced subgraphs of G on $n - k$ vertices are mutually nonisomorphic. In other words, G has property A_k means that, if X, Y are two distinct k-subsets of $V(G)$, then $G - X \not\cong G - Y$. It is easy to show (Exercise 2.5) that if G has property A_{k+1}, then it also has property A_k and that if it has property A_1, then it is asymmetric. We shall show that, for any fixed k, a.e. graph has property A_k.

Lemma 2.5 *Let $W \subseteq V$, $|W| = t$, $|V| = n$, and let $\rho : W \to V$ be an injective function that is not the identity. Let $g = g(\rho)$ be the number of elements $w \in W$ such that $\rho(w) \neq w$. Then there is a set I_ρ of pairs of (distinct) elements of W, containing at least $2g(t-2)/6$ pairs, such that $I_\rho \cap \rho(I_\rho) = \emptyset$.*

Proof Consider those pairs $v, w \in W$ such that at least one is moved. (All pairs are taken to contain distinct elements.) There are $g(t - g) + \binom{g}{2}$ such pairs. For all but at most $g/2$ of these pairs, $\{v, w\} \neq \{\rho(v), \rho(w)\}$ (the exceptions are when $\rho(v) = w$ and $\rho(w) = v$). Let E_ρ be the set of all such pairs. Then

$$|E_\rho| \geq g(t - g) + \binom{g}{2} - g/2 = g(t - g/2 - 1) \geq g(t/2 - 1).$$

Define a graph H_ρ with vertex-set the pairs in E_ρ and such that each pair $\{v, w\}$ is adjacent to the pair $\{\rho(v), \rho(w)\}$. In H_ρ, all degrees are at most 2. Degrees equal to 1 could arise because $\{\rho(v), \rho(w)\}$ could contain an element not in W, and so the pair would not be in E_ρ. Degrees equal to 2 could arise because $\{v, w\}$ could be adjacent to both $\{\rho(v), \rho(w)\}$ and $\{\rho^{-1}(v), \rho^{-1}(w)\}$.

Therefore the components of H_ρ are isolated vertices, paths or cycles. Let I_ρ be a set of independent (that is, mutually not adjacent) vertices in H_ρ. Therefore, for any pair $\{v, w\} \in I_\rho$, $\{\rho(v), \rho(w)\}$ is not in I_ρ.

Now, all isolated vertices in H_ρ are independent, at least half of the vertices on a path are independent and at least one third of the vertices on a cycle are independent, the extreme case here being a triangle. Therefore

$$|I_\rho| \geq |E_\rho|/3 \geq 2g(t-2)/6,$$

as required. □

Corollary 2.6 *Let* $G \in \mathcal{G}(n, \frac{1}{2})$, $W \subset V = V(G)$ *and* $|W| = t$. *Let* $\rho : W \to V$ *be an injective function that is not the identity. Let* $g = g(\rho)$ *be the number of elements* $w \in W$ *such that* $\rho(w) \neq w$. *Let* S_ρ *be the event*

'ρ *gives an isomorphism from* $G[W]$ *to* $G[\rho(W)]$'.

Then

$$P(S_\rho) \leq \left(\frac{1}{2}\right)^{2g(t-2)/6}.$$

Proof Let I_ρ be the set constructed in the previous lemma. Now, for a given pair $\{v, w\} \in I_\rho$, the event

'$\{v, w\}$ and $\{\rho(v), \rho(w)\}$ are both edges or nonedges'

has probability $1/2$. These events, as they range over all pairs $\{v, w\} \in I_\rho$, are mutually independent because they involve distinct pairs. But S_ρ requires all these events simultaneously. Therefore, by independence,

$$P(S_\rho) \leq \left(\frac{1}{2}\right)^{|I_\rho|} \leq \left(\frac{1}{2}\right)^{2g(t-2)/6},$$

as required. □

The result of this corollary is the crux of the proof of the following theorem: There are too many independent correct 'hits' required for ρ to be an isomorphism, and the probability therefore becomes small as n increases.

Theorem 2.7 (Korshunov; Müller; Bollobás) *Let k be a fixed nonnegative integer and let $G \in \mathcal{G}(n, \frac{1}{2})$. Let p_n denote the probability that*

$$\exists W \subseteq V(G) = V = \{1, 2, \ldots, n\},$$

with $|W| = n - k$ and such that

$$\exists \rho : W \to V, \rho \neq \text{id}, \rho \text{ is an isomorphism from } G[W] \text{ to } G[\rho(W)].$$

Then, $\lim_{n \to \infty} p_n = 0$.

Hence, a.e. graph has property A_k.

Proof Pick a particular $W \subset V$ with $|W| = n - k$. This can be done in $\binom{n}{n-k}$ ways, and

$$\binom{n}{n-k} = \frac{n(n-1)\ldots(n-k+1)}{k!} < n^k.$$

Let $t = n - k$. Let $\rho : W \to V$ be injective and not the identity, and let $g = g(\rho)$ be the number of vertices of W that are moved by ρ. Let S_ρ be the event defined in the previous corollary.

Now, for a given value of g between 1 and t, how many functions ρ are there such that $g(\rho) = g$? Such a function is determined by the set $\{w : \rho(w) \neq w\}$ and by the values it takes on this set. Therefore, there are less than n^{2g} such ρ. Therefore, for a given fixed W, the probability of a nontrivial isomorphism is given by

$$\sum_{\rho \neq \text{id}} P(S_\rho) = \sum_{g=1}^{t} \sum_{\rho : g(\rho) = g} P(S_\rho)$$

$$\leq \sum_{g=1}^{t} n^{2g} \left(\frac{1}{2}\right)^{2g(t-2)/6}$$

$$= \sum_{g=1}^{t} \left[n^2 2^{(2-t)/3}\right]^g$$

$$< \sum_{g=1}^{t} \left[4^{1/3} n^2 2^{-t/3}\right]^g.$$

Now $t = n - k > 12(k + 1) \lg n$ for sufficiently large n. Therefore

$$4^{1/3} n^2 2^{-t/3} < 4^{1/3} n^2 2^{-4(k+1)} \lg n$$
$$= \frac{4^{1/3} n^2}{n^{4(k+1)}}$$
$$\leq \frac{4^{1/3}}{n^{2(k+1)}}$$
$$< \frac{1}{n^{k+1}}$$

where the last inequality follows if $4^{1/3} < n^{k+1}$.

Therefore

$$\sum_{\rho \neq \text{id}} P(S_\rho) < \sum_{g=1}^{t} \left(\frac{1}{n^{k+1}} \right)^g$$
$$< \sum_{g=1}^{n} \left(\frac{1}{n^{k+1}} \right)^g$$
$$= \frac{n^{n(k+1)} - 1}{n^{n(k+1)}(n^{k+1} - 1)}.$$

But all this is for fixed W. Therefore the required probability is

$$p_n < n^k \frac{n^{n(k+1)} - 1}{n^{n(k+1)}(n^{k+1} - 1)},$$

and this tends to 0 as n tends to infinity. $\qquad\square$

2.3 Graph symmetries and the spectrum

The adjacency matrix $A = A(G)$ of a graph with vertex-set

$$\{v_1, v_2, \ldots, v_n\}$$

is an $n \times n$ matrix defined by $A_{ij} = 1$ if vertices v_i and v_j are adjacent and $A_{ij} = 0$ if vertices v_i and v_j are not adjacent. The *characteristic polynomial* of G is $\phi(G; x)$ defined by $\det(xI - A)$. The *spectrum* of G is the spectrum of the adjacency matrix A of G, that is, the roots of the characteristic polynomial, or the eigenvalues of A. Eigenvalues and eigenvectors of $A(G)$ are also referred to as the eigenvalues and eigenvectors of G. Note that the spectrum is invariant under a change of labellings of the vertices of G. Also, since A is symmetric, its eigenvalues are real and it has a complete set of orthogonal eigenvectors.

In this and the next section we shall consider some of the relationships that exist between the spectrum of G and $\text{Aut}(G)$. Let π be a permutation of $V(G)$. Then π can be represented by a permutation matrix P, where $P_{ij} = 1$ if and only if $\pi(v_i) = v_j$ and 0 otherwise. The relationship between automorphisms of G and its adjacency matrix arises mainly through the following simple result.

Lemma 2.8 *Let A be the adjacency matrix of a graph G and let π be a permutation of $V(G)$. Then π is an automorphism of G if and only if $PA = AP$, where P is the permutation matrix representing π.*

Proof Suppose first that $AP = PA$. Consider any pair of vertices v_i, v_j. Let $\pi(v_i) = v_h$ and $\pi(v_j) = v_k$. Then

$$(PA)_{ik} = \sum_t P_{it}A_{tk} = A_{hk}$$

and

$$(AP)_{ik} = \sum_s A_{is}P_{sk} = A_{ij}.$$

But $(PA)_{ik} = (AP)_{ik}$, therefore $A_{ij} = A_{hk}$, that is, both are either 1 or 0. This means that either both $\{v_i, v_j\}$ and $\{v_h, v_k\}$ are edges or both are nonedges. Hence π is an automorphism.

Conversely, suppose π is an automorphism. Consider the entries $(AP)_{ik}$ and $(PA)_{ik}$ for $i, k \in \{1, 2, \ldots, n\}$, and which we require to be equal. Let $v_h = \pi(v_i)$ and $\pi(v_j) = v_k$. Then, as earlier,

$$(AP)_{ik} = A_{ij}$$

and

$$(PA)_{ik} = A_{hk}.$$

But π maps $\{v_i, v_j\}$ into $\{v_h, v_k\}$ and π is an automorphism. Therefore these two pairs are either both edges or both nonedges, in other words, $A_{ij} = A_{hk}$. Therefore $(AP)_{ik} = (PA)_{ik}$, as required. $\qquad\square$

Corollary 2.9 *Let x be an eigenvector of A corresponding to the eigenvalue λ and P a permutation matrix corresponding to an automorphism of G. Then Px is also an eigenvector corresponding to λ.*

Proof Since A and P commute,

$$A(Px) = P(Ax) = P(\lambda x) = \lambda(Px),$$

as required. $\qquad\square$

2.4 Simple eigenvalues

Because x and Px could be linearly independent, one important consequence of the last result in the previous section is that nontrivial automorphisms could give rise to eigenvalues with multiplicity greater than 1. This fact will be exploited in this section.

Lemma 2.10 *Let λ be a simple eigenvalue of A and let x be a corresponding eigenvector with real components. If P is a matrix representing an automorphism of G, then $Px = \pm x$.*

Proof Because λ is simple and x and Px are both eigenvectors corresponding to it, then they are linearly dependent. That is, $Px = kx$ for some $k \in \mathbb{R}$ (because x has real components). But, because P is a permutation matrix, P^t is the identity matrix for some $t \in \mathbb{N}$. Therefore $k = \pm \sqrt[t]{1} = \pm 1$. $\qquad\square$

Theorem 2.11 (Mowshowitz; Petersdorf and Sachs) *If all eigenvalues of G are simple, then every nontrivial automorphism of G has order 2. That is, $\mathrm{Aut}(G)$ is an elementary abelian 2-group.*

Proof Let P be any permutation matrix corresponding to an automorphism of G, and let $\{x_1, x_2, \ldots, x_n\}$ be a complete set of eigenvectors of G. Because $Px_i = \pm x_i$ for each i, then $P^2 x_i = x_i$. But because the x_i form a basis, $P^2 = I$, the identity matrix. $\qquad\square$

Theorem 2.12 (Petersdorf and Sachs) *Let G be a vertex-transitive graph that has degree d, and let λ be a simple eigenvalue of G. If $|V(G)|$ is odd, then $\lambda = d$. If $|V(G)|$ is even, then λ takes one of the values $2r - d$ where r is an integer such that $(0 \leq r \leq d)$.*

Proof Let x be a real eigenvector corresponding to the simple eigenvalue λ. Let v_i, v_j be two vertices of G, and let π be an automorphism of G mapping v_i to v_j (such an automorphism must exist because G is vertex-transitive). Let P be the permutation matrix corresponding to π.

Therefore, $(Px)_i = x_j$. But $Px = \pm x$, therefore $x_i = \pm x_j$. By vertex-transitivity, this holds for all $i, j \in \{1, 2 \ldots, n = |V(G)|\}$; that is, the entries x_i all have the same modulus, differing possibly only in sign.

Let $u = (1, 1, \ldots, 1)^{\mathrm{T}}$. Note that d is an eigenvalue of G with eigenvector u. We now consider two cases.

Case 1: n is odd. If $\lambda \neq d$, then the inner product $\langle u, x \rangle = 0$, since eigenvectors corresponding to distinct eigenvalues of the symmetric matrix A are orthogonal. That is, $\sum_1^n x_i = 0$. Because $|x_i| = |x_j|$, the x_i, would have to be paired off as negatives and positives in order to sum to 0. But this is impossible with n odd. Therefore $\lambda = d$.

Case 2: n is even. Let $v_i \in G$. Note that $(Ax)_i = \sum x_j$, where the sum is taken over all j such that v_j is adjacent to v_i. But $(Ax)_i = \lambda x_i$, therefore $\lambda x_i = \sum x_j$, where the sum is as described earlier.

Now, of the x_j in this sum, suppose r are equal to x_i and $d - r$ are equal to $-x_i$. Therefore $rx_i - (d - r)x_i = \lambda x_i$, that is, $\lambda = 2r - d$. $\qquad\square$

2.5 Higher symmetry conditions

In this book we shall be mostly interested in the three types of graph symmetry that we have defined and discussed, namely, vertex-transitivity, edge-transitivity and arc-transitivity. However, in order to put these properties of graphs into a better perspective we shall define, in this section, some extensions of these concepts of graph symmetry. We shall do this rather briefly, first because a more detailed treatment would send us off the track that we would like to follow in this book, and second because there are already several good books, to which we shall refer, that cover these topics quite exhaustively.

First let us start with the idea of *t*-arc-transitivity. A *t-arc* in a graph is a sequence of edges $(e_1, e_2, \ldots, e_t)$ such that, if $1 \leq i \leq t$, $e_i = \{a, b\}$ and $e_{i+1} = \{c, d\}$, then $b = c$. We do not require that the edges in a *t*-arc (even consecutive ones) be distinct.

A graph G is said to be *t-arc-transitive* if, given any two *t*-arcs

$$(e_1, \ldots, e_t) \text{ and } (f_1, \ldots, f_t)$$

in G, there is an automorphism α of G such that

$$(\alpha(e_1), \ldots, \alpha(e_t)) = (f_1, \ldots, f_t).$$

This is a generalisation of arc-transitivity because 1-arc-transitivity reduces to our previous definition of arc-transitivity.

Tutte introduced the concept of *t*-arc-transitivity, and a very good treatment is given in Biggs' book [24]. We limit ourselves to stating the following results without proof. These are long and difficult proofs, and the results can be considered to be amongst the pillars of algebraic graph theory. Biggs' book [24] gives the full proof of Tutte's Theorem.

Theorem 2.13 (Tutte) *There are no finite t-arc-transitive cubic graphs with* $t > 5$.

Theorem 2.14 (Weiss) *Except for cycles, there are no finite t-arc-transitive graphs for* $t \geq 8$.

Another extension of vertex-transitivity and edge-transitivity is distance-transitivity. A graph G is said to be *distance-transitive* if, given any vertices a, b, c, d such that $d(a, b) = d(c, d)$, there is an automorphism of G such that $\alpha(a) = c$ and $\alpha(b) = d$.

It turns out that distance-transitivity is a strong condition that places strict restrictions on the structure of a graph. One of the first major results illustrating and fully describing this strong restriction for cubic graphs was the following.

Theorem 2.15 (Biggs and Smith) *Up to isomorphism, there are exactly twelve finite distance-transitive graphs with degree* 3.

To understand some of the ideas involved in the study of distance-transitive graphs, let us first consider the following. Suppose u, v are vertices in a connected graph G with $d(u, v) = j$. The vertices adjacent to u can be partitioned into three sets: the set A of vertices whose distance from v is j; the set B of vertices whose distance from v is $j + 1$; and the set C of vertices whose distance from v is $j - 1$.

Now, if G is distance-transitive, the numbers $|A|, |B|, |C|$ do not depend on the vertices u, v but only on their distance j. In this case, let these numbers be denoted by $|A| = a_j, |B| = b_j$ and $|C| = c_j$. If the diameter of G is d, then we can write down these numbers as follows:

$$
\iota(G) = \begin{bmatrix} * & c_1 & c_2 & \ldots & c_{d-1} & c_d \\ a_0 & a_1 & a_2 & \ldots & a_{d-1} & a_d \\ b_0 & b_1 & b_2 & \ldots & b_{d-1} & * \end{bmatrix},
$$

and $\iota(G)$ is then called the *intersection array* of G. Because a distance-transitive graph is regular, say of degree k, then $b_0 = k$; also, $a_0 = 0$ and $c_1 = 1$. Moreover, each column of the array sums to k; therefore it is sufficient to give just two of the three rows. The intersection array is therefore often presented as

$$
\iota(G) = [k, b_1, \ldots, b_{d-1}; 1, c_2, \ldots, c_d].
$$

Now, since G is distance-transitive, we can also define what are called the *intersection numbers* of G. Let u, v be any two vertices at distance j from each other. Then the intersection number s_{hij} is defined as the number of vertices that are distance h from u and distance i from v. Clearly, s_{hij} does not depend on the choice of u and v but only on their distance. What is not so trivial is that the $(d+1)^3$ intersection numbers (d is the diameter of G) can be obtained knowing only the entries in the intersection array of G. What is more remarkable is that this result is also true if we relax the condition of distance-transitivity but retain the combinatorial regularity imposed by the intersection array.

Thus, a *distance-regular graph* is a regular connected graph with degree k and diameter d for which the following holds. There are numbers

$$b_0 = k, \quad b_1, \ldots, b_{d-1}, \quad c_1 = 1, \quad c_2, \ldots, c_d$$

such that for each pair (u, v) of vertices with $d(u, v) = j$ we have

(i) the number of vertices adjacent to u and distance $j - i$ from v is c_j $(1 \leq j \leq d)$;

(ii) the number of vertices adjacent to u and distance $j + 1$ from v is b_j $(0 \leq j \leq d - 1)$.

The array $[k, b_1, \ldots, b_{d-1}; 1, c_2, \ldots, c_d]$ is then called the intersection array of the distance-regular graph.

Distance-transitive graphs are clearly distance-regular, but the converse is not true. Quite remarkably, for a distance-regular graph the intersection numbers can be defined, and they can be determined from the intersection array. Details of how this is done are given in [24], where the important question of determining the feasibility of intersection arrays (that is, when does there exist a distance-regular graph with a given intersection array?) is also investigated.

Finally, we want to mention one other generalisation of these regularity properties of graphs. This is in the direction of the generalisation from distance-transitivity to distance-regularity, that is, combinatorial regularities that are implied by symmetries of the automorphism group are used as the defining property of the class of graphs in question. Thus, let G be a connected graph with diameter d, and let F be a function $F : \{1, \ldots, d\} \to \mathbb{N}$ such that any pair of vertices at distance i from each other have exactly $F(i)$ shortest paths joining them. Such a graph is called an *F-geodetic graph*. If $F(i) = 1$ for all i, then G is simply called *geodetic*.

Clearly, distance-transitive and distance-regular graphs are F-geodetic, but the converse is far from true. An F-geodetic graph need not even be regular, as seen by trees, for example, which are geodetic. However, combined with other properties, more can be said about F-geodetic graphs. We present here, without proof, two such results.

Theorem 2.16 (Scapellato) *A bipartite F-geodetic graph is regular if and only if it is distance-regular.*

Theorem 2.17 (Scapellato) *A bipartite F-geodetic graph with diameter 4 is* (i) *a tree, or* (ii) *a distance-regular graph, or* (iii) *the graph associated with a balanced incomplete block design with* $\lambda = 1$.

2.6 Exercises

2.1 Show that an asymmetric graph must have at least six vertices and that the graph shown in Figure 2.1 has the least number of edges amongst all asymmetric graphs on six vertices. Show also that the smallest asymmetric tree has seven vertices.

2.2 Let G be a vertex-transitive graph and let $\Gamma = \mathrm{Aut}(G)$. Show that G is arc-transitive if and only if the stabiliser Γ_v of any vertex v is transitive on the neighbours of v.

2.3 Let the graph G have odd order and suppose that it is regular of degree at least 1. Suppose that G is edge-transitive. Show that it is also vertex-transitive.

2.4 Let G be regular of degree $d \geq \frac{|V(G)|}{2}$. Suppose that G is edge-transitive. Show that it is also vertex-transitive.

2.5 We have already said that the smallest semisymmetric graph has order 20 and degree 4. It is called the Folkman Graph [77], and it is shown in Figure 2.4. It is defined as follows. Its vertices are the triples (a, b, c) where $a, b \in \mathbf{Z}_2$ and $c \in \mathbf{Z}_5$. The edges of the graph are either of the type $\{(0, b, c), (1, b', c)\}$ or $\{(0, 0, c), (1, b', c+1)\}$ or $(\{(1, 1, c), (1, b', c+2)\}$, for all possible b, c, b', c'. Use a computer to construct this graph and verify that it is semisymmetric.

2.6 The Heawood Graph is a cubic bipartite graph on 14 vertices and 21 edges. It is 4-arc-transitive, and it is the smallest cubic graph of girth 6. It is also distance-transitive. *Sage* has the Heawood Graph implemented, and it can be called up with the command

```
h:=graphs.HeawoodGraph()
```

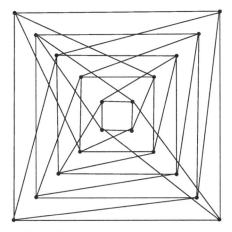

Figure 2.4. The Folkman Graph

Use *Sage* to investigate and verify some of the properties of the Heawood Graph.

2.7 The Ljubljana Graph is a semisymmetric cubic graph on 112 vertices. It is the third smallest cubic semisymmetric graph after the Gray Graph and the Iofinova-Ivanov Graph (which has 110 vertices). The Ljubljana Graph is also implemented in *Sage* and can be called up with

```
l:= graphs.LjubljanaGraph()
```

Use *Sage* to investigate and verify some of the properties of the Ljubljana Graph.

2.8 Show that if G has property A_{k+1}, then it also has property A_k and that if it has property A_1, then it is asymmetric. Hence a.e. graph is asymmetric, which we may call Property A_0. Find an asymmetric graph which has a pair of pseudosimilar vertices. This shows that A_0 does not imply A_1.

2.9 What is the probability that a graph in $\mathcal{G}(n,p)$ has exactly m edges for fixed m between 0 and $\binom{n}{2}$? What is the expected number of edges of a graph in $\mathcal{G}(n,p)$? What is the expected number of triangles?

2.10 Show that the probability that $G \in \mathcal{G}(n,p)$ has $k \leq n$ independent vertices is at most $\binom{n}{k}(1-p)^{\binom{k}{2}}$ and the probability that it contains the complete graph K_k is at most $\binom{n}{k}p^{\binom{k}{2}}$. Show that, for $p = 1/2$, $k \geq 3$ and $n \leq 2^{k/2}$, these probabilities sum to less than 1, and deduce that the Ramsey number $r(k,k)$ is greater than $2^{k/2}$. (See [30] for the definition of Ramsey numbers.)

2.11 Let $G \in \mathcal{G}(n,\frac{1}{2})$ and let $W \subset V(G)$ with $|W| > 50 \ln n$. Show that the probability that there is some other subgraph of G isomorphic to $G[W]$ is $o(1)$. Let I_W be the indicator random variable for there being no other subgraph isomorphic to $G[W]$; therefore, the expected value of I_W is $1 - o(1)$. Let $X = \sum_{|W| > 50 \ln n} I_W$. Show that the expected value of X is $2^n(1 - o(1))$ and that therefore there is some graph $G \in \mathcal{G}(n,\frac{1}{2})$ with $X > 2^n(1 - o(1))$; that is, some G has $2^n(1 - o(1))$ different subgraphs.

[*Note:* By $o(1)$ is meant a real valued function f such that

$$\lim_{n \to \infty} f(n) = 0.]$$

2.12 Let g_n denote the number of nonisomorphic (unlabelled) graphs on n vertices. It can be shown that $g_n \sim 2^{\binom{n}{2}}/n!$ (see, for example, [29] or [103]). Using this result, and the fact that a graph G can be labelled in $|\text{Aut}(G)|/n!$ ways, prove that almost all unlabelled graphs have the trivial automorphism groups and that a property $\mathcal{P}$ holds for almost every graph in $\mathcal{G}(n,p)$ if and only if it holds for almost every unlabelled graph.

2.13 Let G be a graph with adjacency matrix A. Show that $(A^k)_{ij}$ equals the number of walks of length k from vertex v_i to vertex v_j.

2.14 Suppose that G is a regular graph of degree d. Show that d is an eigenvalue of G with eigenvector $(1, 1, \ldots, 1)^t$. Suppose also that G is connected. Show that the multiplicity of the eigenvalue d is 1 and that, for any eigenvalue λ of G, $|\lambda| \leq d$.

2.15 Let G be an arc-transitive graph with degree d and let λ be a simple eigenvalue of G. Prove that $\lambda = \pm d$.

2.16 Find the eigenvalues of the complete graph K_n and the cycle C_n.

2.17 Let the graph G have n vertices, m edges and eigenvalues $\lambda_1, \lambda_2, \ldots, \lambda_n$. Prove that:

(a) $\sum_i \lambda_i = 0$;

(b) $\sum_i \lambda_i^2 = 2m$;

(c) $\sum_i \lambda_i^3 = 6t$, where t is the number of triangles in G;

(d) for any eigenvalue λ_i, $|\lambda_i| \le (2m(n-1)/n)^{\frac{1}{2}}$.

2.18 Let $\pi \in \mathrm{Aut}(G)$ have s cycles of odd length and t cycles of even length when written as a product of disjoint cycles. Show that the number of simple eigenvalues of G is at most $s + 2t$.

2.19 The aim of this exercise is to show that if $\mathrm{Aut}(G)$ is nontrivial, then the characteristic polynomial of G is reducible over $\mathbb{Z}$.

(a) Let $\mathcal{O}_1, \ldots, \mathcal{O}_r$ be the orbits of $V(G)$ under the action of $\mathrm{Aut}(G)$. Show that if $v \in \mathcal{O}_i$, then the number of vertices in $\mathcal{O}_j$ adjacent to v depends only on $\mathcal{O}_i$ and $\mathcal{O}_j$, and not on the choice of vertex v. Denote this number by r_{ij}, and let R be the $r \times r$ matrix (r_{ij}).

(b) Suppose that the vertices of G are labelled $v_1, v_2, \ldots, v_n$ such that first come the elements of $\mathcal{O}_1$, followed by the elements of $\mathcal{O}_2$ and so on. Let A be the corresponding adjacency matrix of G. Let $|\mathcal{O}_i| = n_i$. Suppose that λ is an eigenvalue of R and let $x = (x_1, \ldots, x_r)^{\mathrm{t}}$ be a corresponding eigenvector of λ. Let x^* be the n-vector obtained from x by consecutively repeating n_i times each x_i. Prove that x^* is an eigenvector of A with eigenvalue λ.

(c) Deduce that the characteristic polynomial of R divides that of A.

(d) Deduce that if $\mathrm{Aut}(G)$ is nontrivial, then the characteristic polynomial of G is reducible over $\mathbb{Z}$.

2.20 Show that if G is a t-arc-transitive graph whose vertices have degree at least 3 and whose girth is g, then

$$t \le \frac{1}{2}(g+2).$$

2.21 Show that the automorphism group of a cubic t-arc-transitive graph G acts regularly on the set of t-arcs of G.

2.22 For any vertex v in a connected graph G with diameter d, and, for $1 \le i \le d$, let $G_i(v)$ be the set of vertices that are distance i from v. Prove that G is distance-transitive if and only if it is vertex-transitive and the stabiliser $\mathrm{Aut}(G)_v$ is transitive on the set $G_i(v)$ for each $i \in \{1, 2, \ldots, d\}$ and each $v \in G$.

2.23 Obtain the intersection arrays of K_n, $K_{p,p}$, the line graph $L(K_n)$ for $n \ge 4$, and the cube Q_3. (The hypercube Q_n is the graph whose vertices are binary n-tuples and in which two vertices are adjacent if and only if they differ in exactly one entry (the graph Q_3 is called the cube).)

2.24 Let u, v be at distance k in a graph G. For every $i \in \mathbb{N}$, let $N_i(u, v)$ denote the set of all $w \in G$ such that $d(u, w) = i$ and $d(w, v) = k - i$. Show that G is F-geodetic if and only if $|N_1(u, v)|$ depends only on $d(u, v)$. Show also that, in this case, $F(k) = c_0 c_1 \ldots c_k$ for each k.

2.7 Notes and guide to references

Tutte's Theorem on the degree of $\frac{1}{2}$-arc-transitive graphs first appeared in [250]. A full treatment of random graphs is given in [29]. Theorem 2.7 has been proved independently in [27, 132, 196]. The proof that we give follows mostly [4]; Exercise 2.8 is also from [4]. For the relationship between properties of almost all unlabelled graphs and of almost all labelled graphs see [29, 103].

An excellent reference for more information on the graph spectrum is [24]. Reference [58] gives a comprehensive treatment of the topic. Theorem 2.11 first appeared in [194] and [214], and Theorem 2.12 first appeared in [214]. The result in Exercise 2.19 is from [195].

The classic text [24] gives an extensive treatment of the results mentioned here on t-arc-transitivity, distance-transitivity and distance-regularity. The last two topics in particular are given an encyclopaedic coverage (up to 1989) in [40]. Tutte's Theorem on cubic t-arc-transitive graphs first appeared in [249]. Weiss' result appeared in [255], and it makes essential use of the classification theorem of group theory. The theorem of Biggs and Smith appeared in [25]. Using the classification of finite simple groups, it can now also be shown that for any given $k > 2$ there are only finitely many finite distance-transitive graphs with degree k (see [49]).

The study of F-geodetic graphs probably appeared in [53]. The theorems of Scapellato discussed earlier discussed earlier can be found in [228] and [229], respectively.

A study of some of the many fascinating properties of the Ljubljana Graph can be found in [56, 124].

3

Cayley Graphs

3.1 Cayley colour graphs

Let Γ be an abstract group and X a set of generators for Γ. The *Cayley colour graph* of Γ with respect to X, denoted by $\mathrm{Col}(\Gamma, X)$, is in fact a directed graph with labels (colours) on its arcs. It is defined as follows. Its vertex-set is equal to Γ and its arc-set is

$$\{(\alpha, \alpha\sigma) : \alpha \in \Gamma, \sigma \in X\}.$$

Moreover, the arc $(\alpha, \alpha\sigma)$ is given the colour σ.

In other words, the vertices α, β are joined by an arc from α to β that is coloured σ if and only if there is some element σ in X such that $\alpha\sigma = \beta$. Therefore the arc-set can also be written as

$$\{(\alpha, \beta) : \alpha, \beta \in \Gamma, \alpha^{-1}\beta \in X\}.$$

The arc (α, β) with $\alpha^{-1}\beta \in X$ is therefore given the colour $\alpha^{-1}\beta$.

The following facts about $\mathrm{Col}(\Gamma, X)$ are easily verified:

- $1 \in X$ if and only if every vertex of $\mathrm{Col}(\Gamma, X)$ is incident to a loop coloured 1; for this reason we often tacitly assume that $1 \notin X$.
- The out-degree and the in-degree of every vertex in $\mathrm{Col}(\Gamma, X)$ is equal to $|X|$.
- Any directed walk in $\mathrm{Col}(\Gamma, X)$ consisting of edges $e_1, e_2, \ldots, e_k$ coloured $\sigma_1, \sigma_2, \ldots, \sigma_k$, respectively, corresponds to the element $\sigma_1\sigma_2 \ldots \sigma_k \in \Gamma$. In fact, if α is joined to β by such a walk, then $\beta = \alpha\sigma_1\sigma_2 \ldots \sigma_k$. Also, a cycle in $\mathrm{Col}(\Gamma, X)$ corresponds in this fashion to a sequence of elements of Γ whose product is equal to 1.
- X generates Γ if and only if $\mathrm{Col}(\Gamma, X)$ is strongly connected.
- If both σ and σ^{-1} are in X, then for any arc (α, β) coloured σ there is in $\mathrm{Col}(\Gamma, X)$ a corresponding arc (β, α) coloured σ^{-1}. If $\sigma^2 = 1$, then this

situation is often represented by drawing a single edge coloured σ joining α and β.

Examples

1. Consider the symmetric group S_3 that is isomorphic to the dihedral group D_3. It has a presentation

$$S_3 = \langle \alpha, \beta | \alpha^2 = \beta^3 = (\alpha\beta)^2 = 1 \rangle.$$

(For example, take $\alpha = (12)$ and $\beta = (123)$.) Then the Cayley colour graph of S_3 with respect to $X = \{\alpha, \beta\}$ is shown in Figure 3.1. (In this figure, an undirected edge represents a pair of oppositely directed arcs joining the same pair of vertices and both are coloured α. The directed arcs in the figure are those coloured β.)

2. The dihedral group D_4 has a presentation

$$D_4 = \langle \alpha, \beta | \alpha^2 = \beta^4 = (\alpha\beta)^2 = 1 \rangle.$$

(Take $\alpha = (13)$ and $\beta = (1234)$.) The corresponding Cayley colour graph of D_4 is shown in Figure 3.2. The undirected edges and the arcs in this figure are coloured as in the previous example.

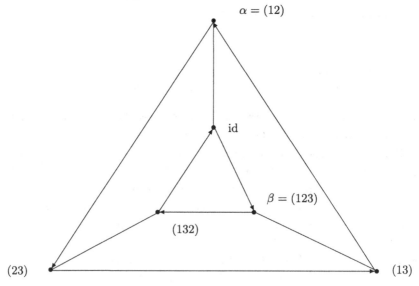

Figure 3.1. A Cayley colour graph of S_3

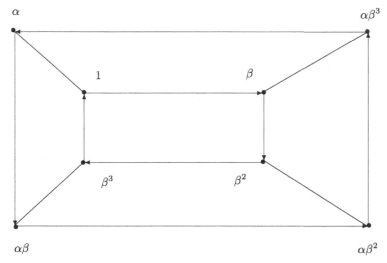

Figure 3.2. A Cayley colour graph of D_4

A *colour preserving automorphism* of $\mathrm{Col}(\Gamma, X)$ is a permutation π of $V(\mathrm{Col}(\Gamma, X)) = \Gamma$ such that, for all $\alpha, \beta \in \Gamma$, (α, β) is an arc coloured σ if and only if $(\pi(\alpha), \pi(\beta))$ is also an arc coloured σ.

The set of all colour preserving automorphisms of $\mathrm{Col}(\Gamma, X)$ forms a group under composition of functions, and it is denoted by

$$\mathrm{Aut}(\mathrm{Col}(\Gamma, X)).$$

Lemma 3.1 *Let Γ be a nontrivial finite group with a generating set X. Let π be a permutation of $V(\mathrm{Col}(\Gamma, X))$. Then π is a colour preserving automorphism of $\mathrm{Col}(\Gamma, X)$ if and only if $\pi(\alpha\sigma) = \pi(\alpha)\sigma$ for all $\alpha \in \Gamma$ and $\sigma \in X$.*

Proof Let π be a colour preserving automorphism, and let $\alpha \in \Gamma$ and $\sigma \in X$. Because the arc $(\alpha, \alpha\sigma)$ is coloured σ, the arc $(\pi(\alpha), \pi(\alpha\sigma))$ is also coloured σ. Therefore $\pi(\alpha\sigma) = \pi(\alpha)\sigma$.

Conversely, suppose $\pi(\alpha\sigma) = \pi(\alpha)\sigma$ for all $\alpha \in \Gamma$ and $\sigma \in X$. Now, an arc (α, β) is coloured σ if and only if $\beta = \alpha\sigma$, that is, if and only if $\pi(\beta) = \pi(\alpha\sigma) = \pi(\alpha)\sigma$, and hence if and only if $(\pi(\alpha), \pi(\beta))$ is an arc coloured σ. Therefore π is a colour preserving automorphism. $\square$

Theorem 3.2 *Let Γ be a nontrivial finite group with a generating set X. Then the action of $\mathrm{Aut}(\mathrm{Col}(\Gamma, X))$ on $V(\mathrm{Col}(\Gamma, X))$ is equivalent to the permutation* *group* $(L(\Gamma), \Gamma)$, *that is, the left regular action of* Γ *on itself. In particular,* $\mathrm{Aut}(\mathrm{Col}(\Gamma, X)) \simeq \Gamma$.

Proof For any $\alpha \in \Gamma$ recall that λ_α denotes the permutation of Γ given by $\lambda_\alpha : \beta \mapsto \alpha\beta$. We need to show that $(L(\Gamma), \Gamma)$ is equivalent to $(\mathrm{Aut}(\mathrm{Col}(\Gamma, X)), \Gamma)$. We therefore must first show that $\lambda_\alpha : \beta \mapsto \alpha\beta$ is, in fact, a colour preserving automorphism of $\mathrm{Col}(\Gamma, X)$. But this is so because the pair (β, γ) is an arc coloured σ if and only if $\gamma = \beta\sigma$, that is, $\alpha\gamma = \alpha\beta\sigma$, and hence $\lambda_\alpha(\gamma) = \lambda_\alpha(\beta)\sigma$; and this is so if and only if $(\lambda_\alpha(\beta), \lambda_\alpha(\gamma))$ is an arc coloured σ.

Therefore, the only thing left to prove is that any colour preserving automorphism of $\mathrm{Col}(\Gamma, X)$ is, in fact, an element of $L(\Gamma)$. Suppose that $\phi \in \mathrm{Aut}(\mathrm{Col}(\Gamma, X))$ and suppose that $\phi(1) = \alpha$. We claim that $\phi = \lambda_\alpha$.

For, any $\beta \in \Gamma$ can be written as $\beta = \sigma_1\sigma_2\ldots\sigma_t$, with the $\sigma_i \in X$ (we are not excluding the possibility that consecutive σ_i are equal). Therefore, $\phi(\beta) = \phi(1\beta) = \phi(1\sigma_1\sigma_2\ldots\sigma_t)$. Applying the previous lemma successively gives that $\phi(\beta) = \phi(1)\sigma_1\sigma_2\ldots\sigma_t = \alpha\beta = \lambda_\alpha(\beta)$, as required. $\square$

3.2 Frucht's and Bouwer's Theorems

It is natural to wonder if, given a group Γ, there is a graph having Γ as its full automorphism group. The answer to this question depends on whether Γ is given as an abstract group or a permutation group.

Theorem 3.3 (Frucht) *Let* Γ *be a finite group. Then there exists a finite graph* G *such that* $\mathrm{Aut}(G) \simeq \Gamma$.

Proof Let $X = \{\sigma_1, \sigma_2, \ldots, \sigma_t\}$ be any generating set of Γ and construct $\mathrm{Col}(\Gamma, X)$. Then Γ is isomorphic to the group of colour preserving automorphisms of $\mathrm{Col}(\Gamma, X)$. What we must do now is transform $\mathrm{Col}(\Gamma, X)$ into a graph G such that no new automorphisms are formed.

It is actually quite a simple matter to encode the directions of the arcs and their colours as edges. Every arc coloured σ_i is replaced by a 'gadget' as shown in Figure 3.3.

It is clear that every automorphism of $\mathrm{Col}(\Gamma, X)$ induces an automorphism of G, and vice versa. This is because none of the new vertices can be mapped into any of the old ones by an automorphism and only gadgets of the same type can be mapped into each other, and this can be done only in the 'proper' direction corresponding to the orientation of the two arcs. Therefore, $\mathrm{Aut}(G) \simeq \Gamma$, as required. $\square$

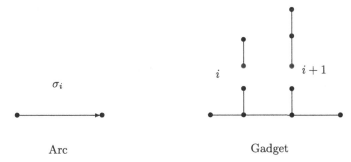

Arc   Gadget

Figure 3.3. Replacing arcs by gadgets

It is important to stop a while and see what we have *not* proved by means of Frucht's Theorem.

Frucht's Theorem only tells us that, given a group Γ, there is a graph G whose automorphism group is abstractly isomorphic to Γ. It does not tell us that, given a permutation group Γ acting on a set Y, there is a graph G such that the action of $\mathrm{Aut}(G)$ on $V(G)$ is equivalent to the action of Γ on Y. In fact, it is quite easy to show that there are permutation groups Γ for which there is no graph whose automorphism group is equivalent to Γ (see exercises at the end of this chapter).

The next theorem, however, tells us that, given a permutation group Γ acting on a set Y, one can, at least, always construct a graph G such that its automorphism group restricted to an induced subgraph of G is equivalent to (Γ, Y).

Theorem 3.4 (Bouwer) *Let (Γ, Y) be a group of permutations of a set Y. Then there exists a graph G such that $\mathrm{Aut}(G) \simeq \Gamma$, X is a nonempty subset of $V(G)$, X is invariant under the action of $\mathrm{Aut}(G)$ and the action of $\mathrm{Aut}(G)$ on X is equivalent to the action of Γ on Y.*

[*Remark:* In the following proof, the subgraph H induced by X will, in fact, be the null graph. Therefore, H alone would admit all of S_X as a group of automorphisms. The rest of G is there, so to speak, in order to 'kill off' some of these permutations, leaving only those in Γ.]

Proof Let G' be a graph constructed as in the proof of Frucht's Theorem such that $\mathrm{Aut}(G') \simeq \Gamma$. Recall that $\Gamma \subseteq V(G')$. Recall also that G' contains two types of vertices, those which we can call 'gadget vertices', as the ones shown in Figure 3.3, and those elements which are actually elements of the group Γ considered as vertices of G'. To help us distinguish between the roles of the elements of Γ as permutations and as elements of a group and their role as vertices of G' we shall denote the set Γ of vertices of G' by Γ'; similarly, if $\alpha \in \Gamma$, then we shall denote the corresponding vertex of G' by α'. We shall refer to these vertices as the 'group vertices' of G'.

Note that, by Theorem 3.2 and by the construction in Frucht's Theorem, the permutation group $(\mathrm{Aut}(G'), \Gamma')$, that is, the restriction of $\mathrm{Aut}(G')$ to the group vertices of G', is equivalent to $(L(\Gamma), \Gamma)$.

Now, let the orbits of Y under the action of Γ be $\mathcal{O}_1, \ldots, \mathcal{O}_t$. Let us, for the moment, concentrate on the orbit $\mathcal{O}_1$. Let $y \in \mathcal{O}_1$, and let Γ_y be its stabiliser. Of course, $\Gamma_y \subseteq V(G')$; again we shall denote the set of vertices Γ_y by Γ'_y. Consider the group vertices of G' partitioned as cosets of Γ'_y. For each such coset add a new vertex x and join it to all the group vertices in that coset and to no other vertex. Let G'' be the graph G' with the new vertices added in this way.

It is important to ensure that the graph G'' resulting from this construction does not have any new automorphisms that are not in $\mathrm{Aut}(G')$. One crude way to ensure that no new unwanted automorphisms have been introduced by the new vertices is to create a copy of the complete graph K_p (p sufficiently large) for each new vertex x and let x be one of the vertices of the corresponding complete graph. We shall now denote by $\mathrm{Aut}(G'')$ the automorphism group $\mathrm{Aut}(G')$ with its action extended also to the new vertices. Note that $\mathrm{Aut}(G'')$ is still isomorphic to Γ. Moreover, since the action of the permutation group $(\mathrm{Aut}(G'')$ on the group vertices of G' is equivalent to the left translation of Γ on itself, the action of $\mathrm{Aut}(G'')$ on the set of new vertices is equivalent to the permutation group $(L^{\Gamma_y}(\Gamma), \Gamma/\Gamma_y)$. But by Theorem 1.1, this permutation group is equivalent to the action of Γ on Y restricted to the orbit $\mathcal{O}_1$. That is, the action of $\mathrm{Aut}(G'')$ on the new vertices attached to the cosets of Γ_y is equivalent to the action of Γ on $\mathcal{O}_1$.

Repeat this procedure for each of the orbits $\mathcal{O}_2, \ldots, \mathcal{O}_t$. Let X be the totality of all new vertices added and joined to some coset of some stabiliser of Γ. Let G denote the resulting graph. It then follows that G and X have the required properties.   □

3.3 Cayley graphs and digraphs

In this section we shall describe what is one of the most important ways of constructing a graph or digraph from a given group. In effect, this construction is equivalent to that of $\mathrm{Col}(\Gamma, X)$, but with some extra conditions on the set X and without labels on the arcs.

So, let Γ be a finite group and S a *generating* set for Γ such that $1 \notin S$. The *Cayley digraph* $\mathrm{Cay}(\Gamma, S)$ is defined as follows. Its vertex-set is Γ, and two vertices α and β are joined by the arc (α, β) if and only if $\beta = \alpha\sigma$ for some $\sigma \in S$. The set S is often called the *connecting set* of the Cayley digraph.

The condition $1 \notin S$ is imposed so that no arcs of the form (α, α) are present in $\mathrm{Cay}(\Gamma, S)$. Also, S is required to be a generating set of Γ so that $\mathrm{Cay}(\Gamma, S)$ is strongly connected; that is, given any two vertices α and β, there is a directed path from α to β.

An important class of Cayley digraphs occurs if S is required to satisfy a further condition, namely, that, for any σ in S, σ^{-1} is also in S; if $S^{-1} = \{s^{-1} : s \in S\}$, then this property can be denoted by $S^{-1} = S$.

When $S^{-1} = S$, an arc (α, β) is in $\mathrm{Cay}(\Gamma, S)$ if and only if (β, α) is also an arc of $\mathrm{Cay}(\Gamma, S)$. In this case we shall always assume that every pair of oppositely directed arcs $(\alpha, \beta), (\beta, \alpha)$ is replaced by the single edge $\{\alpha, \beta\}$, and the resulting graph $G = \mathrm{Cay}(\Gamma, S)$ is then called a *Cayley graph*. This graph G would therefore have $V(G) = \Gamma$ and

$$\begin{aligned} E(\mathrm{Cay}(\Gamma, S)) &= \{\{\alpha, \alpha\sigma\} : \alpha \in \Gamma, \sigma \in S\} \\ &= \{\{\alpha, \beta\} : \alpha, \beta \in \Gamma, \alpha^{-1}\beta \in S\}. \end{aligned}$$

The property $S^{-1} = S$ ensures that $E(G)$ is this way well defined because, if $\beta = \alpha\sigma, \sigma \in S$, then $\alpha = \beta\sigma^{-1}$ and σ^{-1} is also in S. In fact, replacing each pair of arcs $(\alpha, \beta), (\beta, \alpha)$ by the edge $\{\alpha, \beta\}$ is simply the reverse process of obtaining $\overleftrightarrow{G}$ from a graph G. It is clear that, if $S^{-1} \neq S$, then every vertex in the digraph $\mathrm{Cay}(\Gamma, S)$ has both out-degree and in-degree equal to $|S|$; while if $S^{-1} = S$, then the graph $\mathrm{Cay}(\Gamma, S)$ is regular of degree $|S|$.

The next theorem tells us much more than the fact that all of the vertices of a Cayley graph have degree $|S|$. Since our main interest in this chapter is in undirected Cayley graphs, we shall state this result here for this case. This theorem (which can be considered to be the principal result about Cayley graphs), however, also holds for the general case of Cayley digraphs; the proof is quite similar to the undirected case and is left as an exercise.

Theorem 3.5 *Let $G = \mathrm{Cay}(\Gamma, S)$ be a Cayley graph. Then the permutation group $(L(\Gamma), \Gamma)$ is a subpermutation group of $(\mathrm{Aut}(G), V(G))$. Therefore G is vertex-transitive and $\mathrm{Aut}(G)$ contains a subgroup isomorphic to Γ and acting regularly on $V(G)$. Conversely, a graph G is a Cayley graph $\mathrm{Cay}(\Gamma, S)$ for some Γ and $S \subseteq \Gamma$ if $(\mathrm{Aut}(G), V(G))$ contains a subpermutation group isomorphic to Γ acting regularly on $V(G)$.*

Proof Consider the action of left translation of Γ on itself; that is, for any $\alpha \in \Gamma$, let λ_α be the permutation mapping $\beta \in \Gamma$ to $\alpha\beta$. We need to prove that

λ_α is in $\text{Aut}(G)$. But

$$\{\beta, \gamma\} \in E(G) \Leftrightarrow \gamma = \beta\sigma, \ \sigma \in S$$
$$\Leftrightarrow \beta^{-1}\gamma \in S$$
$$\Leftrightarrow \beta^{-1}\alpha^{-1}\alpha\gamma \in S$$
$$\Leftrightarrow (\alpha\beta)^{-1}(\alpha\gamma) \in S$$
$$\Leftrightarrow \{\alpha\beta, \alpha\gamma\} \in E(G)$$
$$\Leftrightarrow \{\lambda_\alpha(\beta), \lambda_\alpha(\gamma)\} \in E(G).$$

Therefore $\lambda_\alpha \in \text{Aut}(G)$, as required.

For the converse, let $\Gamma \leq \text{Aut}(G)$ such that Γ acts regularly on $V(G)$. Let v_0 be a vertex of G, and let $S = \{\sigma \in \Gamma : \sigma(v_0) \text{ adjacent to } v_0\}$. We claim that G is isomorphic to $\text{Cay}(\Gamma, S)$.

Observe first that $S^{-1} = S$ because suppose that $\sigma \in S$ and $\sigma(a) = v_0$, $\sigma(v_0) = b$ (note that $a \neq v_0 \neq b$ because Γ acts regularly on $V(G)$ and σ is not the identity because G does not contain any loops). Therefore b is adjacent to v_0. But $\sigma(av_0) = v_0 b \in E(G)$ and, since σ is an automorphism of G, av_0 must also be an edge of G. Therefore $a = \sigma^{-1}(v_0)$ is adjacent to v, that is, $\sigma^{-1} \in S$, as required.

We now define a function $\phi : V(\text{Cay}(\Gamma, S)) \to V(G)$ by $\phi(\alpha) = \alpha(v_0)$, and we claim that ϕ is an isomorphism between the graphs $\text{Cay}(\Gamma, S)$ and G. The bijectivity of ϕ is easy to establish. To demonstrate that ϕ preserves adjacency, let $\alpha, \beta \in V(\text{Cay}(\Gamma, S)) = \Gamma$. Then,

$$\{\alpha, \beta\} \in E(\text{Cay}(\Gamma, S)) \Leftrightarrow \alpha^{-1}\beta \in S$$
$$\Leftrightarrow \alpha^{-1}\beta = \sigma \quad (\sigma \in S)$$
$$\Leftrightarrow \{v_0, \alpha^{-1}\beta(v_0)\} \in E(G)$$
$$\Leftrightarrow \{\alpha(v_0), \beta(v_0)\} \in E(G)$$
$$\Leftrightarrow \{\phi(\alpha), \phi(\beta)\} \in E(G).$$

Therefore ϕ is an isomorphism, as required.   □

Although $\Gamma \leq \text{Aut}(\text{Cay}(\Gamma, S))$, it is often the case that $\text{Aut}(\text{Cay}(\Gamma, S))$ is larger than Γ (see, for example, Exercise 3.11). The instance when $\Gamma = \text{Aut}(\text{Cay}(\Gamma, S))$ is an important situation that is considered in some more detail in a subsequent chapter.

3.4 The Doyle-Holt Graph

We are now in a position to give an example of a $\frac{1}{2}$-arc-transitive graph which we know by Tutte's Theorem (Theorem 2.4) must have even degree. Holt first published this graph in [107], but it was eventually discovered that Doyle had studied it independently in his Harvard senior dissertation in 1976. It is now known as the Doyle-Holt Graph. We shall here describe it as a Cayley graph. Let Γ be the group

$$\langle \alpha, \beta | \alpha^9 = \beta^3 = 1, \beta^{-1}\alpha\beta = \alpha^4 \rangle$$

of order 27, and let S be the generating set

$$\{\beta\alpha, \beta\alpha^{-1}, \beta^2\alpha^2, \beta^2\alpha^{-2}\}.$$

We shall use the GAP package in order to build this graph (denoted by *dh*) and to analyse the structure of its automorphism group.

```
# Load the package GRAPE
LoadPackage("grape");

# Construct the group from its presentation
f:=FreeGroup(2);

rels:=[f.1^9,f.2^3,
       f.1^f.2*f.1^5];
g:=f/rels;

# Construct the connecting set of the Cayley graph
gens:=[g.2*g.1, g.2*g.1^8, g.2^2*g.1^2, g.1^2*g.2^7];

# Construct the Cayley graph dh
dh:=CayleyGraph(g,gens);

# Since we are studying the action on the edges, construct
# the line-graph of dh so we can then study the action on
# the vertices of the line-graph
ldh:=EdgeGraph(dh);

# Construct the automorphism group of ldh
lgg:=AutGroupGraph(ldh);
```

Having constructed the automorphism group of *ldh* we can ask GRAPE if the action of this group is transitive on the vertices of *ldh*, that is, if *dh* is edge-transitive. GRAPE answers the query

```
IsTransitive(lgg, Vertices(ldh));
```

with 'True'. We can ask for the size of the group with the command

```
Size(lgg);
```

to which GAP returns with the answer 54, which is the number of vertices of *ldh*, that is, the number of edges of *dh*. Therefore the automorphism group of the Doyle-Holt Graph acts regularly on its edges, that is, it cannot act transitively on the arcs. Since we know that it is vertex-transitive (it is a Cayley graph) we have a $\frac{1}{2}$-arc-transitive graph.

To learn more about this remarkable graph, one should read Holt's short paper [107] and also Doyle's paper [64], both of which give different computer-free proofs that the graph is $\frac{1}{2}$-arc-transitive. Bouwer [34] had earlier constructed $\frac{1}{2}$-transitive graphs, but his smallest had fifty-four vertices. B. Alspach, D. Marušič and L. A. Nowitz [7] eventually showed that there can be no $\frac{1}{2}$-arc-transitive graphs smaller than the Doyle-Holt Graph. The paper [125] illustrates a more sophisticated use of combinatorial computer packages to bring out some remarkable connections between the Doyle-Holt Graph and other topics in algebraic combinatorics.

3.5 Non-Cayley vertex-transitive graphs

Although all Cayley graphs are vertex-transitive, it is not the case that Cayley graphs characterise vertex-transitive graphs. The most frequently cited example of a vertex-transitive graph that is not isomorphic to any Cayley graph is the Petersen graph P, and we now briefly sketch a proof that P is not a Cayley graph. Recall first that P has order 10, it is cubic and its diameter is 2. Now suppose that P were isomorphic to a Cayley graph $\text{Cay}(\Gamma, S)$. Then Γ would have to be a group of order 10 and the size of S would have to be 3. Also, the diameter of $\text{Cay}(\Gamma, S)$ is the least positive integer d such that $\Gamma = S \cup S^2 \cup \cdots \cup S^d$, where $S^{i+1} = \{\sigma\tau : \sigma \in S, \tau \in S^i\}$. Now, there are only two nonisomorphic groups of order 10, namely $\mathbb{Z}_{10}$ and D_5, and, checking all 3-sets S in each with the property that $1 \notin S$ and $S^{-1} = S$, one finds that each gives a diameter greater than 2.

Some results that we have already obtained can be brought together to give a construction of a whole family of vertex-transitive graphs that are not Cayley graphs.

Theorem 3.6 (Watkins) *Let G be a graph on at least five vertices that is edge-transitive but not bipartite and whose vertices have odd degree. Let $H = L(G)$ be the line-graph of G. Then H is a vertex-transitive graph that is not a Cayley graph.*

Proof Clearly H is vertex-transitive because G is edge-transitive. Suppose H is also a Cayley graph. Then by Theorem 3.5 there is a subgraph $\mathcal{H}$ of $\mathrm{Aut}(H)$ that acts regularly on $V(H)$.

Note that $\mathcal{H}$ consists of edge-automorphisms of G. But by Whitney's Theorem, since G has at least five vertices, any edge-automorphism of G is induced by an automorphism of G. Therefore $\mathcal{H}$ is induced by a subgraph of $\mathrm{Aut}(G)$ acting regularly on the edges of G. Therefore the action of $\mathcal{H}$ on G is edge-transitive but not arc-transitive.

But G is not bipartite. Therefore the action of $\mathcal{H}$ must be transitive on $V(G)$. Hence, by Tutte's Theorem, the degree of G must be even, which is a contradiction.

Therefore H cannot be a Cayley graph. $\qquad\qquad\square$

A simple example of a graph on at least five vertices that is edge-transitive but not bipartite and whose vertices have odd degree, as required in the statement of the theorem, is given by the complete graph on $2n$ vertices. The line-graph of K_{2n} is therefore a vertex-transitive non-Cayley graph.

3.6 Coset graphs and Sabidussi's Theorem

We shall now define a construction which is more general than Cayley graphs and which will characterise vertex-transitive graphs.

Let Γ be a group, $\mathcal{H} \leq \Gamma$ and S a subset of Γ such that (i) $S \subseteq \Gamma - \mathcal{H}$, (ii) $S^{-1} = S$ and (iii) $\mathcal{H} \cup S$ generates Γ. The *coset graph* $\mathrm{Cos}(\Gamma, \mathcal{H}, S)$ is defined as the graph whose vertices are the left cosets of $\mathcal{H}$ in Γ and in which $\alpha\mathcal{H}, \beta\mathcal{H}$ are adjacent if and only if $\alpha^{-1}\beta \in \mathcal{H}S\mathcal{H}$.

It is useful to note the following points about this definition:

- If $\mathcal{H} = \{1\}$, then $\mathrm{Cos}(\Gamma, \mathcal{H}, S)$ becomes the Cayley graph $\mathrm{Cay}(\Gamma, S)$.

- Condition (i) implies that there are no loops (note that $1 \notin S$ follows from this condition) and condition (ii) means that all edges are undirected. Condition (iii) gives that $\mathrm{Cos}(\Gamma, \mathcal{H}, S)$ is connected.
- It is easy to check the well-definition of $\mathrm{Cos}(\Gamma, \mathcal{H}, S)$. Let $\alpha\mathcal{H} = \gamma\mathcal{H}$ and $\beta\mathcal{H} = \delta\mathcal{H}$. Therefore $\gamma = \alpha\epsilon$ and $\delta = \beta\phi$, for ϵ and ϕ in $\mathcal{H}$. Therefore

$$\gamma^{-1}\delta \in \mathcal{H}S\mathcal{H} \Leftrightarrow \gamma^{-1}\delta = \epsilon_1\sigma\phi_1, (\epsilon_1, \phi_1 \in \mathcal{H}, \sigma \in S)$$
$$\Leftrightarrow \epsilon^{-1}\alpha^{-1}\beta\phi = \epsilon_1\sigma\phi_1$$
$$\Leftrightarrow \alpha^{-1}\beta = (\epsilon\epsilon_1)\sigma(\phi_1\phi^{-1}) \in \mathcal{H}S\mathcal{H}.$$

- An equivalent way to describe adjacency in $\mathrm{Cos}(\Gamma, \mathcal{H}, S)$ is by saying that two cosets $\alpha\mathcal{H}, \beta\mathcal{H}$ are adjacent if and only if there exist $\gamma \in \alpha\mathcal{H}$ and $\delta \in \beta\mathcal{H}$ such that $\gamma^{-1}\delta \in S$.

Theorem 3.7 *The coset graph* $\mathrm{Cos}(\Gamma, \mathcal{H}, S)$ *is vertex-transitive.*

Proof Let $G = \mathrm{Cos}(\Gamma, \mathcal{H}, S)$. For each $\alpha \in \Gamma$ define $\lambda_\alpha : V(G) \to V(G)$ by $\lambda_\alpha(\beta\mathcal{H}) = \alpha\beta\mathcal{H}$. It is easy to check that λ_α is a permutation. Also, it is an automorphism of G because $\beta\mathcal{H}, \gamma\mathcal{H}$ are adjacent if and only if $\beta^{-1}\gamma \in \mathcal{H}S\mathcal{H}$, and this holds if and only if $(\alpha\beta)^{-1}(\alpha\gamma) \in \mathcal{H}S\mathcal{H}$, that is, if and only if $\alpha\beta\mathcal{H}$ and $\alpha\gamma\mathcal{H}$ are adjacent.

Therefore G is vertex-transitive because, if $\beta\mathcal{H}, \gamma\mathcal{H}$ are two vertices of G, then $\lambda_{\gamma\beta^{-1}}(\beta\mathcal{H}) = \gamma\mathcal{H}$. $\qquad\square$

We can now show that any vertex-transitive graph is a coset graph.

Theorem 3.8 (Sabidussi) *Let G be a vertex-transitive graph. Then G is isomorphic to some coset graph* $\mathrm{Cos}(\Gamma, \mathcal{H}, S)$.

Proof Let $\Gamma = \mathrm{Aut}(G)$ and let v be a vertex of G. Let $\mathcal{H} = \Gamma_v$ be the stabiliser of v. Let $S = \{\sigma \in \Gamma : \sigma(v) \text{ is adjacent to } v\}$. We now show that G and $\mathrm{Cos}(\Gamma, \mathcal{H}, S)$ are isomorphic. Note first that $S \cap \mathcal{H} = \emptyset$. Also, $S^{-1} = S$ (the proof of this is very similar to the analogous statement in the proof of Theorem 3.5).

Now define a function $\phi : V(\mathrm{Cos}(\Gamma, \mathcal{H}, S)) \to V(G)$ by $\phi(\alpha\mathcal{H}) = \alpha(v)$. We claim that ϕ is an isomorphism from $\mathrm{Cos}(\Gamma, \mathcal{H}, S)$ to G.

The well-definition of ϕ and its bijectivity are easy to show. Let us demonstrate that ϕ preserves adjacency. Let $\alpha\mathcal{H}, \beta\mathcal{H}$ be two cosets. Then,

$$\{\alpha\mathcal{H}, \beta\mathcal{H}\} \in E(\mathrm{Cos}(\Gamma, \mathcal{H}, S)) \Leftrightarrow \alpha^{-1}\beta \in \mathcal{H}S\mathcal{H}$$
$$\Leftrightarrow \gamma^{-1}\alpha^{-1}\beta\delta^{-1} = \sigma \ (\gamma, \delta \in \mathcal{H}, \ \sigma \in S)$$

$$\Leftrightarrow \{v, \gamma^{-1}\alpha^{-1}\beta\delta^{-1}(v)\} \in E(G)$$
$$\Leftrightarrow \{v, \alpha^{-1}\beta(v)\} \in E(G)$$
$$\Leftrightarrow \{\alpha(v), \beta(v)\} \in E(G)$$
$$\Leftrightarrow \{\phi(\alpha\mathcal{H}), \phi(\beta\mathcal{H})\} \in E(G).$$

$\square$

3.7 Double coset graphs and semisymmetric graphs

Cayley graphs do not manage to represent all transitive graphs because some transitive graphs do not have a subgroup of the automorphism group acting regularly on their vertex-sets, that is, with a trivial stabiliser for any vertex. Coset graphs manage to overcome this problem by basing their construction on the cosets of the stabiliser so that they are working on a vertex-set which includes the stabiliser $v = \mathcal{H}$ itself as a single vertex. Therefore, in the left action of the group on v, the stabiliser of v is $\mathcal{H}$ itself, as required.

This construction provides the idea for representing edge-transitive but not vertex-transitive graphs by using cosets of two subgraphs of the automorphism group, each one being a stabiliser of a vertex from one of the two colour classes, respectively. First we need some definitions.

Let Γ be a group and $\mathcal{H}, \mathcal{K}$ two subgroups of Γ. The *double coset graph* $\text{Cos}(\Gamma, \mathcal{H}, \mathcal{K})$ is defined as the graph whose vertices are the left cosets of $\mathcal{H}$ and $\mathcal{K}$ in Γ and in which $\alpha\mathcal{H}, \beta\mathcal{K}$ are adjacent if and only if their intersection is nonempty.

The proof of the following result is left as an exercise.

Theorem 3.9 *Let Γ be a finite group and $\mathcal{H}, \mathcal{K}$ subgroups of Γ whose union generates the group. Then the graph $\text{Cos}(\Gamma, \mathcal{H}, \mathcal{K})$ is a connected edge-transitive bipartite graph with the two sets of cosets of $\mathcal{H}$ and $\mathcal{K}$ being the colour classes of $\text{Cos}(\Gamma, \mathcal{H}, \mathcal{K})$ whose vertices have degrees $|\mathcal{H}|/|\mathcal{H} \cap \mathcal{K}|$ and $|\mathcal{K}|/|\mathcal{H} \cap \mathcal{K}|$, respectively.*

Conversely, let G be a graph on which the group Γ acts edge-transitively but not vertex-transitively. Let uv be an edge of G and $\mathcal{H}$ and $\mathcal{K}$ the stabilisers in Γ of u and v, respectively. Then the union of $\mathcal{H}$ and $\mathcal{K}$ generates Γ and G is isomorphic to $\text{Cos}(\Gamma, \mathcal{H}, \mathcal{K})$.

This result can be helpful in constructing semisymmetric graphs by considering $\text{Cos}(\Gamma, \mathcal{H}, \mathcal{K})$ with $\mathcal{H}, \mathcal{K}$ having the same order. In fact, we can construct the Gray Graph this way. We let α be the permutation (1 2 3) and β the permutation (1 4 7)(2 5 8)(3 6 9). Let $\Gamma = \langle \alpha, \beta \rangle$, $\mathcal{H} = \langle \alpha \rangle$ and $\mathcal{K} = \langle \beta \rangle$. We shall this time use GAP to construct the coset graph. First, the groups are constructed with the following commands. The last two commands create the two sets of cosets.

```
gap> a:=(1,2,3);
gap> b:=(1,4,7)(2,5,8)(3,6,9);
gap> gamma:=Group(a,b);
gap> h:=Subgroup(gamma, [a]);
gap> k:=Subgroup(gamma, [b]);
gap> hcoset:=RightCoset(h,a);
gap> kcoset:=RightCoset(k,b);
```

Then the GRAPE package is loaded with

```
gap> LoadPackage("grape");
```

The next command uses GRAPE's 'Graph' function to construct the coset graph. In its simplest form, this function is similar to Sage's 'DiGraph()' or 'Graph()' in that you give it a set of vertices and a boolean function which will be evaluated at all pairs of the vertex-set and will make them adjacent if it returns the value 'True'. But in GRAPE the function also takes a group which acts as a group of automorphisms of the constructed graph. In this case we need to specify that the action is indicated by the option 'OnRight', which means group multiplication on the right cosets.

```
gray:= Graph(gamma,[hcoset,kcoset], OnRight, function(x,y)
   return x<>y and Intersection(x,y)<>[]); end);
```

The following sequence of commands and GAP's responses confirms that the graph gray is a cubic graph on 54 vertices which is edge-transitive but not vertex-transitive, and it is therefore the Gray Graph.

```
gap> OrderGraph(gray)
54
gap> VertexDegrees(gray)
[ 3 ]
gap> grp1:=AutGroupGraph(gray);
gap> IsTransitive(grp1,Vertices(gray));
false
gap> lgray:=EdgeGraph(gray) #construct the line graph of gray;
gap> # lgray is the line-graph of gray
```

```
gap> grp2:=AutGroupGraph(lgray);
gap> IsTransitive(grp2,Vertices(lgray));
true
```

We shall have occasion, in a later chapter, to use coset graphs again for another application.

3.8 Hamiltonicity

Lovász has conjectured that all vertex-transitive graphs have a Hamiltonian path. It has also been conjectured that all Cayley graphs have a Hamiltonian cycle. We shall show in this section that Cayley graphs of abelian groups do, in fact, have a Hamiltonian cycle. This well-known result is included here because it illustrates nicely the parallels between computations in the group Γ and graph theoretic properties in $\text{Cay}(\Gamma, S)$.

First we need to establish some notation. Let $\sigma = (\sigma_1, \sigma_2, \ldots, \sigma_r)$ be a sequence of elements $\sigma_i \in S \subseteq \Gamma$. Then $\pi_i(\sigma)$ will denote the i-th partial product $\sigma_1 \sigma_2 \ldots \sigma_i$, and, if $r \geq 3$, $\hat{\sigma}$ will denote the sequence $(\sigma_2, \ldots, \sigma_{r-1})$. Given two sequences $\sigma = (\sigma_1, \sigma_2, \ldots, \sigma_r)$ and $\tau = (\tau_1, \tau_2, \ldots, \tau_q)$, then (σ, τ) will denote the sequence

$$(\sigma_1, \sigma_2, \ldots, \sigma_r, \tau_1, \tau_2, \ldots, \tau_q).$$

The sequence σ^n is defined recursively by $\sigma^1 = \sigma$ and $\sigma^n = (\sigma, \sigma^{n-1})$. Also, σ^{-1} will denote the sequence $(\sigma_r^{-1}, \sigma_{r-1}^{-1}, \ldots, \sigma_1^{-1})$.

It is clear that a Hamiltonian cycle in a Cayley graph $\text{Cay}(\Gamma, S)$ corresponds to a sequence $\sigma = (\sigma_1, \sigma_2, \ldots, \sigma_r)$, where each term σ_i is in S, $r = |\Gamma|$, all partial products $\pi_i(\sigma)$ are distinct and $\pi_r(\sigma) = 1$. We shall therefore call such a sequence of elements of Γ a *Hamiltonian sequence of Γ in S*. Therefore, proving that $\text{Cay}(\Gamma, S)$ is Hamiltonian is equivalent to showing that Γ has a Hamiltonian sequence in S.

Lemma 3.10 *Let Γ be an abelian group. Let $T \subseteq \Gamma$ with $T^{-1} = T$ and such that $\mathcal{H} = \langle T \rangle$. Let $\alpha \notin T$ and let $S = T \cup \{\alpha, \alpha^{-1}\}$. Let $\mathcal{K} = \langle S \rangle$.*

Then a Hamiltonian sequence of $\mathcal{H}$ in T can be extended to a Hamiltonian sequence of $\mathcal{K}$ in S.

Proof Clearly we may assume that $\mathcal{K}$ is strictly larger than $\mathcal{H}$; otherwise, any Hamiltonian sequence of $\mathcal{H}$ in T is also a Hamiltonian sequence of $\mathcal{K}$ in S. Let $\tau = (\tau_1, \tau_2, \ldots, \tau_r)$ be a Hamiltonian sequence of $\mathcal{H}$ in T. Suppose that j

is the least positive integer such that $\alpha^j \in \mathcal{H}$ (therefore $j > 1$). Then clearly $|\mathcal{K}| = |\mathcal{H}| \cdot j = rj$.

Consider first the case when j is odd. Let σ be the sequence

$$(\alpha, \hat{\tau}, \alpha, \hat{\tau}^{-1}).$$

Then it can easily be checked that

$$(\sigma^{(j-1)/2}, \tau, (\alpha^{-1})^{j-1})$$

is a Hamiltonian sequence of $\mathcal{K}$ in S.

Now consider the case when j is even. Again let σ be as earlier. Then the sequence

$$(\sigma^{(j-2)/2}, \alpha, \hat{\tau}, \tau_1, \hat{\tau}^{-1}, \tau_1^{-1}, (\alpha^{-1})^{j-1})$$

is also a Hamiltonian sequence of $\mathcal{K}$ in S. $\square$

Using this lemma recursively, the following result follows.

Corollary 3.11 *Let Γ be an abelian group. Let $S \subseteq \Gamma$ with $S^{-1} = S$ and such that $\Gamma = \langle S \rangle$. Let $\mathcal{H}$ be a subgroup of Γ generated by $T = T^{-1} \subset S$. Then a Hamiltonian sequence of $\mathcal{H}$ in T can be extended to a Hamiltonian sequence of Γ in S.*

Theorem 3.12 *A Cayley graph $\mathrm{Cay}(\Gamma, S)$ of an abelian group Γ with at least three vertices contains a Hamiltonian cycle.*

Proof Suppose first that S contains an element α of order n at least 3. Let $\mathcal{H}$ be the subgroup generated by α. Clearly, the sequence $(\alpha)^n$ is a Hamiltonian sequence of $\mathcal{H}$ in $\{\alpha\}$. By Corollary 3.11, this Hamiltonian sequence can be extended to a Hamiltonian sequence of Γ in S.

Now suppose that S contains no element of order at least 3. Since $|\Gamma| \geq 3$, S must contain at least two elements β, γ of order 2. Let $\mathcal{H}$ be the group generated by these two elements. Clearly $(\beta, \gamma, \beta, \gamma)$ is a Hamiltonian sequence of $\mathcal{H}$ in $\{\alpha, \beta\}$. Therefore, by Corollary 3.11, this sequence can be extended to a Hamiltonian sequence of Γ in S. $\square$

While the question of the Hamiltonicity of Cayley graphs is still open, it is very easy to construct non-Hamiltonian Cayley digraphs (see Exercise 3.18).

Table 3.1. *Character table of the cyclic group of order* 6 *where* ω *is a primitive sixth root of unity.*

g	1	a	a^2	a^3	a^4	a^5
$\chi_1(g)$	1	1	1	1	1	1
$\chi_2(g)$	1	ω	ω^2	ω^3	ω^4	ω^5
$\chi_3(g)$	1	ω^2	ω^4	1	ω^2	ω^4
$\chi_4(g)$	1	ω^3	1	ω^3	1	ω^3
$\chi_5(g)$	1	ω^4	ω^2	1	ω^4	ω^2
$\chi_6(g)$	1	ω^5	ω^4	ω^3	ω^2	ω

3.9 Characters of abelian groups and Cayley graphs

The theory of group characters makes possible the study of the relationship between the group Γ and the spectrum of the Cayley graph $\mathrm{Cay}(\Gamma, S)$. By restricting ourselves to abelian groups we can give a short but self-contained treatment of a typical result in this area. Therefore in this section any group mentioned is assumed to be finite and abelian.

A character of an abelian group Γ is a homomorphism π from Γ to the complex field $\mathbb{C}$. Some properties of characters follow easily from the definition. For example, let $g \in \Gamma$ have order k. Then, since π is a homomorphism, $\pi(g^k) = 1$, therefore $\pi(g)$ is equal to a kth root of unity. Also, again since π is a homomorphism, $\pi(g)\pi(g^{-1}) = 1$, therefore $\pi(g^{-1}) = (\pi(g))^{-1} = \overline{\pi(g)} = \pi(\bar{g})$. It can be shown that an abelian group of order n has exactly n characters. Table 3.1 gives the characters of the cyclic group of order 6.

Such a table is called the *character table* of the group. Using the characterisation of abelian groups as direct products of cyclic groups makes it easy to construct the character table of abelian groups in general using the following result.

Theorem 3.13 *If* χ, ψ *are characters of the abelian groups* Γ_1 *and* Γ_2, *respectively, then* $\chi \times \psi$ *defined by*

$$(\chi \times \psi)(\alpha, \beta) = \chi(\alpha)\psi(\beta) \qquad (\alpha \in \Gamma_1, \beta \in \Gamma_2)$$

is a character of the direct product $\Gamma_1 \times \Gamma_2$.

The following theorem then makes it possible to determine the spectrum of Cayley graphs of abelian groups without having to work with the adjacency matrix of the graph. Some simple examples of the use of this result are given in Exercises 3.22, 3.23, 3.24 and 3.25.

Theorem 3.14 *Let Γ be an abelian group and $S \subseteq \Gamma$. Let $G = Cay(\Gamma, S)$. Let the elements of Γ be ordered as $g_1, g_2, \ldots, g_n$, and let A be the adjacency matrix of G with respect to this ordering. Suppose that the characters of G are $\chi_1, \chi_2, \ldots, \chi_n$. Then the eigenvalues $\lambda_1, \lambda_2, \ldots, \lambda_n$ of A are given by*

$$\lambda_i = \sum_{s \in S} \chi(s)$$

with corresponding eigenvector

$$\chi_i = (\chi_i(g_1), \chi_i(g_2), \ldots, \chi_i(g_n))^T.$$

Proof Consider the k-th entry of $A\chi_i$.

$$\begin{aligned}
(A\chi_i)_k &= \sum_{j=1}^{n} A_{kj} \chi_i(g_j) \\
&= \sum_{g_j : g_k \sim g_j} \chi_i(g_j) \\
&= \sum_{s \in S} \chi_i(g_k \cdot s) \\
&= \sum_{s \in S} \chi_i(g_k) \chi_i(s) \\
&= \chi_i(g_k) \sum_{s \in S} \chi_i(s) \\
&= (A\chi_i)_k \cdot \sum_{s \in S} \chi_i(s)
\end{aligned}$$

$\square$

3.10 Growth rates

In this book we are concerned with finite graphs or digraphs, and hence finite automorphism groups. However, we shall make a very brief foray in this section into the world of infinite graphs and groups in order to give a flavour of how some of the graph theoretic ideas that we are discussing are extended to the infinite case. We must emphasise that this is only a very brief excursion, because any treatment that is more lengthy would lead us too far astray from the main scope of this book. The interested reader must consult at least the works that we shall be referring to in order to obtain a clearer picture of the topics that we shall be hinting at.

We shall actually be mentioning two measures of growth of infinite groups. First, let Γ be a finitely generated infinite group. Let S be a set of generators of Γ. The growth rate of Γ measures the rate at which successive powers of S cover the group. This can be conveniently worded in the terminology of graphs and Cayley graphs.

Thus, let G be an infinite graph (that is, one whose vertex-set and edge-set are both infinite) but which is *locally finite*, that is, all degrees of its vertices are finite. Let G be connected and v a vertex of G, and let the *ball* of radius n about v, denoted by $B(n, v)$, be the set of vertices at distance at most n from v. The graph G is said to have *polynomial growth of degree d* if there exist constants c_1 and c_2 such that

$$c_1 n^d \leq |B(n, v)| \leq c_2 n^d.$$

It has *subexponential growth* if

$$\liminf_{n \to \infty} \frac{|B(n, v)|}{a^n} = 0$$

for any $a > 1$, and *exponential growth* otherwise.

Now, if $G = \text{Cay}(\Gamma, S)$ is a locally finite infinite Cayley graph for a finitely generated infinite group Γ, then Γ is said to have polynomial, subexponential or exponential growth if G has. It is easily seen that this definition is independent of the choice of generating set S.

The following are two important results in this field. First a definition. A group is said to be *virtually nilpotent* if it has a nilpotent subgroup of finite index.

Theorem 3.15 (Gromov) *A group has polynomial growth if and only if it is virtually nilpotent.*

Theorem 3.16 (Trofimov) *Let G be a vertex-transitive (infinite) graph. Then the following are equivalent.*

(i) *G has polynomial growth.*

(ii) *$V(G)$ under the action of $\text{Aut}(G)$ admits a system of imprimitivity σ with finite equivalent classes such that $\text{Aut}(G/\sigma)$ is finitely generated, virtually nilpotent and the stabiliser of G/σ in the automorphism group of G/σ is finite.*

Now let us consider another notion of the rate of growth of an infinite group, this time given as a permutation group. Let (Γ, Y) be a permutation group where both Γ and Y are infinite. Then $f_n(\Gamma)$ denotes the number of orbits of Γ

on the set of n-element subsets of Y, and $F_n(\Gamma)$ denotes the number of orbits of Γ on the set of ordered n-tuples of distinct elements of Y.

The infinite permutation group (Γ, Y) is called an *oligomorphic permutation group* if $f_n(\Gamma)$ is finite for all $n \in \mathbb{N}$. This is equivalent to saying that $F_n(\Gamma)$ is finite for all $n \in \mathbb{N}$ because the following connection between $f_n(\Gamma)$ and $F_n(\Gamma)$ is easily seen to hold for oligomorphic groups (and its proof is left as an exercise for the reader):

$$f_n(\Gamma) \leq F_n(\Gamma) \leq n! f_n(\Gamma).$$

For oligomorphic permutation groups both the sequences $\langle f_n(\Gamma) \rangle$ and $\langle F_n(\Gamma) \rangle$ are monotonic nondecreasing (this result is easily seen for $F_n(\Gamma)$, but it is much deeper for $f_n(\Gamma)$). The main problem in oligomorphic groups is to determine how these sequences behave and, in particular, whether they grow rapidly and whether they grow smoothly.

One of the most striking results in this field is the following theorem. Here, an oligomorphic permutation group is said to be highly set-transitive if it is k-homogeneous for all $k \in \mathbb{N}$.

Theorem 3.17 (Macpherson) *Let (Γ, Y) be an oligomorphic permutation group. Then there is a constant $c > 1$ such that, if Γ is primitive but not highly set-transitive, then $f_n(\Gamma) \geq c^n$ for all sufficiently large n.*

For general permutation groups, the following theorem holds. Part (a) is due to Macpherson and part (b) to Pouzet.

Theorem 3.18 (Macpherson; Pouzet) *(a) Let (Γ, Y) be an infinite permutation group. Then either there are some $k \in \mathbb{N}$ and constants $a, b > 0$ such that $an^k \leq f_n(\Gamma) \leq bn^k$ for all n, or else the sequence $\langle f_n(\Gamma) \rangle$ grows faster than any polynomial in n.*
(b) In the latter case, $f_n(\Gamma) \geq \exp(n^{1/2-\epsilon})$ for all $n \geq n_0(\epsilon)$.

3.11 Exercises

3.1 Suppose that in the definition of a Cayley (di)graph the set S is not required to be a set of generators of Γ. Show that the resulting (di)graph is disconnected with each component being an isomorphic copy of $\mathrm{Cay}(\langle S \rangle, S)$.

3.2 The alternating group of degree 4, A_4, has a presentation

$$A_4 = \langle \alpha, \beta \,|\, \alpha^2 = \beta^3 = (\alpha\beta)^3 = 1 \rangle.$$

(For example, take $\alpha = (12)(34)$ and $\beta = (123)$.) Obtain the Cayley colour graph of A_4 with respect to the set $\{\alpha, \beta\}$.

3.3 The direct product $\mathbb{Z}_2 \times \mathbb{Z}_4$ has a presentation

$$\mathbb{Z}_2 \times \mathbb{Z}_4 = \langle \alpha, \beta | \alpha^2 = \beta^4 = 1, \alpha\beta = \beta\alpha \rangle.$$

Obtain the corresponding Cayley colour graph with respect to the set $\{\alpha, \beta\}$. Compare this with the Cayley graph of D_4 shown in Figure 3.2.

3.4 Modify the gadgets in the proof of Frucht's Theorem so that the graph constructed has minimum degree greater than 1.

3.5 This exercise is intended to illustrate Bouwer's Theorem. As in Exercise 1.11, consider the action of the alternating group A_4 on the set $X = \{1, 2, 3, 4\}$ and let $\mathcal{H}$ be the stabiliser of the element 1 under this action. First draw the Frucht Graph G for A_4 starting from the Cayley colour graph produced in Exercise 3.2. Now, identify the four cosets of $\mathcal{H}$ as subsets of the vertices of G. Add four new vertices $v_1, \ldots, v_4$ joining each one to all the vertices in exactly one of the cosets, respectively. Verify that the automorphism group of the resulting graph is still isomorphic to A_4 and that its action on $v_1, \ldots, v_4$ is equivalent to the action of A_4 on Y.

3.6 This exercise provides a simple example of a permutation group which is not equivalent to an automorphism of a graph G acting on $V(G)$.

(a) (Kagno [118]) Let $V(G) = \{1, 2, \ldots, n\}$ and suppose $\alpha : i \mapsto i + 1 \mod n$ is an automorphism of G (that is, $\alpha = (1\ 2 \ldots n)$). Show that $\beta = (1\ n)(2\ n - 1)(3\ n - 2) \ldots$ is also an automorphism of G.
   Deduce that if Γ is the cyclic group acting on $N = \{1, 2, \ldots, n\}$ generated by the permutation $(12 \ldots n)$, then there is no graph G such that $(\mathrm{Aut}(G), V(G)) \equiv (\Gamma, N)$. Obtain a graph G such that $\mathrm{Aut}(G) \simeq \Gamma$.

(b) Show by an example that this result is false for digraphs. At what point does your proof break down in the case of a digraph?

(c) Kagno's result can be slightly generalised and proved this way.
   Suppose G is a graph such that Γ is an abelian group of automorphism of G acting regularly on $V(G)$. Therefore $G = \mathrm{Cay}(\Gamma, S)$ for some $S \subseteq \Gamma$. Show that there is an automorphism of Γ which fixes S and therefore Γ cannot be the full automorphism group of G.

3.7 Let $V(G) = \{1, 2, \ldots, n\}$ and suppose $\mathrm{Aut}(G)$ is abelian. Show that the permutation $\alpha = (12 \ldots n) \notin \mathrm{Aut}(G)$.

3.8 Show that there is no graph G whose automorphism group is equivalent to the alternating group A_n.

3.9 Show that $f_n(\Gamma) \leq F_n(\Gamma) \leq n! f_n(\Gamma)$ for an oligomorphic permutation group Γ.

3.10 Formulate and prove a result analogous to Theorem 3.5 for Cayley digraphs.

3.11 In this exercise we give two examples of Cayley graphs $\mathrm{Cay}(\Gamma, S)$ with automorphism groups larger than Γ.

(a) Let Γ be a finite cyclic group generated by the element g and let $S = \{g, g^{-1}\}$. Show that $\mathrm{Aut}(\mathrm{Cay}(\Gamma, S))$ is larger than Γ.

(b) Let Γ be the dihedral group defined by

$$\langle \alpha, \beta | \alpha^2 = \beta^6 = 1, \beta\alpha = \alpha\beta^{-1} \rangle,$$

and let $G = \mathrm{Cay}(\Gamma, S)$, where $S = \{\alpha, \alpha\beta, \alpha\beta^3\}$. Show that $\mathrm{Aut}(G)$ is larger than Γ.

3.12 Let $G = \text{Cay}(\Gamma, S)$ and let $\alpha \in \Gamma$. Suppose λ_α is the left regular action of α on $V(G)$, and let $\mathcal{O}_1, \mathcal{O}_2, \ldots, \mathcal{O}_k$ be the orbits in $V(G)$ under the action of λ_α and all its powers. Show that these orbits have the same size and, in fact, that they are the right cosets of the cyclic group generated by α.

3.13 Let the vertex-transitive graph G with automorphism group Γ be isomorphic to $\text{Cos}(\Gamma, \mathcal{H}, S)$, as constructed in Theorem 3.8. Show that G is edge-transitive if and only if $\mathcal{H}S\mathcal{H} = \mathcal{H}\{\mu, \mu^{-1}\}\mathcal{H}$ for some $\mu \in \Gamma - \mathcal{H}$.

   Show also that G is arc-transitive if and only if $\mathcal{H}S\mathcal{H} = \mathcal{H}\{\mu\}\mathcal{H}$ for some $\mu \in \Gamma - \mathcal{H}$ such that $\mu = \mu^{-1}$.

3.14 Let G be a Cayley graph $\text{Cay}(\Gamma, S)$ and let $S^* = S \cup \{1\}$. Show that G is a line-graph if and only if $S^* = S_1 \cup S_2$ such that $S_1 \cap S_2 = \{1\}$ and either both S_1 and S_2 are subgroups of Γ or $S_1 = \mathcal{H} \cup \mathcal{H}\alpha$ and $S_2 = \alpha^{-1}\mathcal{H}\alpha \cup \alpha^{-1}\mathcal{H}$, where $\mathcal{H}$ is a subgroup of Γ and $\alpha \in \Gamma$ such that $\mathcal{H} \cap \alpha^{-1}\mathcal{H}\alpha = \{1\}$. [Use the Krausz characterisation of line-graphs (see [22]).]

3.15 Let G be a connected vertex-transitive graph on p vertices where p is an odd prime. Using the Orbit-Stabiliser Theorem, show that $\text{Aut}(G)$ has an element α of order p. Let v_1 be a vertex of G and let v_2 be a neighbour of v_1. Show that α or some power of α maps v_1 to v_2.

   Deduce that G is Hamiltonian.

   Show also that, if G is as above but is a digraph and $\text{Aut}(G)$ is abelian, then G is Hamiltonian.

3.16 Show that a connected vertex-transitive graph G with degree at least 3 is 3-connected (see [253]). Deduce that any two longest cycles in G have at least three vertices in common.

   Now let t be the length of a longest cycle in G and let X be the set of t-subsets of $V(G)$ whose elements are vertices of a longest cycle. Suppose that each vertex of G is in exactly s elements of X. Show that $t|X| = |(V(G)|s$ and $3|X| \leq ts$. Hence show that every connected vertex-transitive of order at least 4 has a cycle of length at least $3n^{\frac{1}{2}}$ (see [16]).

3.17 Let Γ be the dihedral group defined by

$$\langle \alpha, \beta | \alpha^2 = \beta^6 = 1, \beta\alpha = \alpha\beta^{-1} \rangle,$$

and let G be the Cayley graph of Γ with connecting set $\{\alpha, \beta, \beta^{-1}\}$. Find a Hamiltonian cycle in G.

3.18 Consider the Cayley digraph $D = \text{Cay}(\Gamma, S)$ where Γ is the additive group $\mathbb{Z}_{12}$ of integers modulo 12, and $S = \{3, 4\}$. Suppose that D had a Hamiltonian cycle C. Show that if C contains the arc $(1, 4)$ then it can only contain arcs of the form $(i, i + 3)$ and that if C contains the arc $(1, 5)$ then it can only contain arcs of the form $(i, i + 4)$.

   Deduce that the Cayley digraph D cannot have a Hamiltonian cycle.

3.19 (a) Complete the missing details in the above proof that the Petersen graph is not a Cayley graph.

   (b) Show that all vertex-transitive graphs on less than ten vertices are Cayley graphs.

3.20 Let G be a graph and let $\mathcal{H}$ be an abelian subgroup of Aut(G) acting transitively on $V(G)$. Show that for any two vertices u, v of G there is an automorphism α in $\mathcal{H}$ such that $\alpha(u) = v$ and $\alpha(v) = u$.

3.21 Let Γ be a finite group defined by the following presentation:

$$\Gamma = \langle a_1, a_2, \ldots, a_d | R_1 = R_2 = \cdots = R_r = 1\rangle,$$

where the R_i are words in the a_j. The aim of this exercise is to show that $r \geq d$.

Let S be the collection of elements $a_1, a_1^{-1}, \ldots, a_d, a_d^{-1}$, with elements of order 2 appearing twice, and construct the Cayley graph Cay(Γ, S). (This is a slightly modified type of Cayley graph because elements a_j of order 2 give double edges and if some $a_j = 1$, we get loops at all vertices.) Note that, for any $a \in V(G)$, each relation gives a closed walk starting and ending at a. These closed walks between them span the cycle space of G. Using the fact that the dimension of this space is $|E(G)| - |V(G)| + 1$, obtain that $r \geq d$.

3.22 (a) Write down the character table of the group $C_2 \times C_2$. Use this to obtain the eigenvalues of the adjacency matrix of the cycle on four vertices.

(b) The complete graph K_n can be considered to be the Cayley graph Cay(C_n, S) where C_n is the cyclic group of order n and S is all of C_n except the identity element. Use this Cayley graph to determine all the eigenvalues of the adjacency matrix of K_n.

(c) Let G be an abelian group and again consider the graph K_n to be the Cayley graph of G with connecting set S containing all the nonidentity elements of G. Using the fact that all eigenvalues λ of K_n except $\lambda = n$ are equal to 1, show that, for any nontrivial character χ of G,

$$\sum_{g \in G} \chi(g) = 1.$$

3.23 The graph of the cube is the Cayley graph of the group $C_2 \times C_4$ with connecting set $S = \{(a, 1), (1, b), (1, b^3)\}$. Find the eigenvalues of the characteristic polynomial of this graph.

3.24 Determine what Cayley graph is depicted in Figure 3.4 and find its eigenvalues.

3.25 Let C_n be the graph consisting of the cycle on n vertices. For what values of n does C_n have 0 as an eigenvalue?

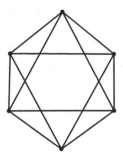

Figure 3.4. The Cayley graph for Exercise 3.24

3.26 (Gromov) Let G be a vertex-transitive infinite graph. Show that

$$|B(n, v)| \cdot |B(5n, v)| \leq |B(4n, v)|^2.$$

[*Hint.* Let Y be a maximal set of vertices in $B(3n, v)$ with minimum distance at least $3n + 1$. Then the disjoint balls $B(n, y)$ for $y \in Y$ are contained in $B(4n, v)$. Also, the balls $B(2n, y)$ for $y \in Y$ cover $B(3n, v)$; therefore, the balls $B(4n, y)$, $y \in Y$, cover $B(5n, v)$.]

3.12 Notes and guide to references

Frucht's Theorem has been strengthened in various ways. For example, it is known that, given any finite group Γ, there is a cubic graph whose automorphism is isomorphic to Γ (see [17, 22, 78]). In the proof of Frucht's Theorem presented earlier no attempt was made to economise on the number of vertices used to construct the graph. Some results on the problem of finding the smallest graphs or digraphs with a given automorphism group can be found in [10, 42, 223]. Bouwer's Theorem can be found in [35].

The Petersen graph is the standard counterexample to many false conjectures in graph theory. The book [108] is devoted to the study of this remarkable graph and its ramifications. Plenty of useful information about special graphs like the Petersen, Gray and Doyle-Holt graphs can be obtained at the site www.mathworld.com and also at www.wolframalpha.com.

Marušič [163] initiated a systematic study of those orders n for which non-Cayley vertex-transitive graphs exist. For more recent results on this question see [81, 182, 183, 233].

Two well-known conjectures about vertex-transitive graphs are Lovász's conjecture [150] that claims that all of them have a Hamiltonian path, and the related conjecture that all Cayley graphs have a Hamiltonian cycle. The fact that Cayley graphs of abelian groups are Hamiltonian is a well-known result. The proof given earlier follows [164]. For a survey on this problem see [261]. Examples of non-Hamiltonian Cayley digraphs are given in [242]. The survey [261] also contains examples that show that vertex-transitive digraphs (even Cayley digraphs) need not even contain a Hamiltonian path.

Exercise 3.21, taken from [244], is a nice application of Cayley graphs to group theory.

The problem of whether a given permutation group is equivalent to an automorphism group of a graph is an important question with implications on issues of computational complexity. For it seems that whether certain group theoretic problems in NP are computationally harder than determining whether two graphs are isomorphic depends on the still open question of whether every permutation group (Γ, Y) is equivalent to the automorphism group of a combinatorial structure whose size is a polynomial in $|Y|$ [106].

Note that even the package [131], which was mentioned in the first chapter, has routines that allow one to work with Cayley graphs and coset graphs.

The notion of the growth rate of a finitely generated group was originally motivated by connections between the curvature of Riemannian manifolds and the growth rate of their fundamental group [187, 186, 262]. The theorems of Gromov and of Trofimov appeared in [93] and [246], respectively. A survey of related topics can be found in [247]. A short treatment is also given in [18]. A good introduction to infinite graphs is the survey paper [254]. A higly recommended and readable introduction to the growth of groups is [184].

Oligomorphic permutation groups are treated in detail in [45], and [49] also contains sections on the growth rate of infinite permutation groups. A good survey paper on such groups is [48]. The two theorems of Macpherson can be found in [157] and [158], while the theorem of Pouzet is in [216]. The proofs of these theorems on the growth rates of groups are all very deep and difficult, and presenting any of them here would be beyond the scope of this book.

4

Orbital Graphs and Strongly Regular Graphs

In this chapter we shall define and explore the most basic properties of what are called orbital graphs. We shall see that, analogously to coset graphs, any vertex-transitive graph is a (generalised) orbital graph. Moreover, this description gives us, in principle, a criterion for arc-transitivity. Consideration of a special type of orbital graph will lead us to strongly regular graphs.

4.1 Definitions and basic properties

We have seen in Chapter 1 that, given a permutation group (Γ, V), there is a natural induced action $(\Gamma, V \times V)$ on the ordered pairs of elements of V. We shall now study this action in more detail. First, let us henceforth, unless stated otherwise, suppose in this chapter that the action of Γ on V is transitive. Then the orbits of Γ on $V \times V$ are called *orbitals*. One of them is $\{(u, u) : u \in V\}$, because (Γ, V) is transitive. We denote this orbital by D_0 and we call it the *trivial orbital*. We usually list the orbitals as $D_0, D_1, \ldots, D_{r-1}$. The number r of orbitals of $(\Gamma, V \times V)$ is called the *rank* of (Γ, V).

The *transpose* A^T of a subset A of $V \times V$ (and, in particular, of any orbital) is defined by $A^T = \{(v, u) : (u, v) \in A\}$. We say that a subset A of $V \times V$, and therefore any orbital, is *self-paired* when $A = A^T$. A self-paired set containing the two arcs (u, v) and (v, u) is considered to be the edge $\{u, v\}$. Note that an orbital D is either self-paired or it contains no pair of opposite arcs. In fact, $D \cap D^T \neq \emptyset$ implies $D = D^T$.

Obviously the rank r of (Γ, V) satisfies $r \geq 2$ unless V has one element. The case $r = 2$ clearly corresponds to a 2-transitive permutation group, and we have already seen that such groups are not very interesting as automorphism groups of graphs. From our point of view, therefore, the first interesting case occurs when $r = 3$. We shall look at this particular case in more detail later

64

in this chapter. We also note that, from now on, any orbital mentioned will be assumed to be nontrivial unless explicitly stated otherwise.

There is a natural way to obtain vertex-transitive graphs or digraphs using orbitals. Thus, let (Γ, V) be a transitive permutation group and let D be an orbital. Then the digraph $G(D)$ whose vertex-set is V and arc-set is D is called an *orbital digraph*. If D happens to be self-paired, then we can, as usual, replace each pair of opposed arcs by an edge and then $G(D)$ is called an *orbital graph*. It is clear that an orbital graph or digraph is not only vertex-transitive but also arc-transitive.

Now, suppose A is a union of orbitals of the permutation group $(\Gamma, V \times V)$. Then the *generalised orbital digraph* $G(A)$ is the digraph with vertex-set V and arc-set A. As usual, if A is self-paired, then we say that the $G(A)$ is a *generalised orbital graph*.

The basic result connecting orbitals with vertex-transitive graphs is the following.

Theorem 4.1 *Let G be a generalised orbital graph or digraph. Then G is vertex-transitive and, moreover, if it is an orbital graph or digraph, then it is also arc-transitive. Conversely, let G be a vertex-transitive graph or digraph. Then G is a generalised orbital graph or digraph and, moreover, if G is arc-transitive, then it is an orbital graph or digraph.*

Proof We have already commented on the first part of the theorem, which follows from the definitions. For the converse, note that the arc-set or edge-set of a graph is invariant under the action of its automorphism group Γ, and, when Γ is transitive, it is therefore a union A of orbitals of (Γ, V). Therefore $G = G(A)$. Moreover, if G is arc-transitive, then A must consist of only one orbital. $\qquad\square$

Examples

1) Consider the natural action of $\Gamma = \langle (12\ldots n)\rangle$ on the set $V = \{1, 2, \ldots, n\}$. The orbitals containing $(1, 2), (1, 3), \ldots, (1, n)$ are all different and hence the rank is n. Each of the corresponding orbital digraphs is a directed cycle of length n or a union of cycles whose length is a fixed divisor of n. These are illustrated in Figure 4.1 for small values of n.

2) As a more interesting example, let $k < n$ and let us consider the action of the symmetric group S_n on the k-subsets of the set $\{1, 2, \ldots, n\}$. For each fixed $i = 0, 1, \ldots, k$ let D_i be the set of all pairs (X, Y), where $X \cap Y$ has cardinality $k - i$. It is easy to check that these are the orbitals of

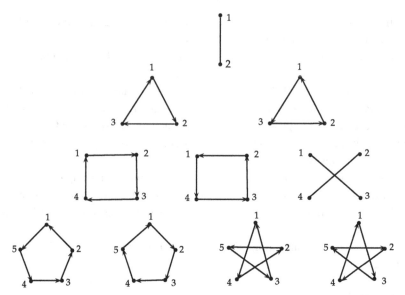

Figure 4.1. Nontrivial orbital graphs and digraphs for the cyclic groups of order $2 \leq n \leq 5$

the given action of S_n (and, of course, D_0 is the trivial orbital, as defined earlier). All the D_i are self-paired for all n and k, with a few exceptions (see Exercise 4.1). The graphs formed by the orbitals D_k are called *Kneser graphs* and are denoted by $K(n,k)$. If $n = 2k+1$, then $K(n,k)$ is also called an *odd graph* and is denoted by O_k. Because they are orbital graphs, the Kneser graphs are all arc-transitive. We shall consider these graphs in more detail in a later chapter.

We shall now present an alternative way of looking at the rank of a transitive permutation group. Thus, let (Γ, V) be a transitive permutation group and let $v \in V$. The orbits of Γ_v acting on V will be called *suborbits* of (Γ, V). Clearly, $\{v\}$ itself is one of these suborbits. We shall show that the number of suborbits is independent of the vertex v chosen and that it is in fact equal to the rank.

Theorem 4.2 *Let (Γ, V) be a transitive permutation group. Then the number of suborbits of (Γ, V) is equal to its rank.*

Proof Let the orbitals of (Γ, V) be, as usual, $D_0, D_1, \ldots, D_{r-1}$.

Let K be the complete graph on the vertex-set V, and let $\overleftrightarrow{K}$ be, as usual, obtained from K by replacing every edge by a pair of opposite arcs. Let the arcs of $\overleftrightarrow{K}$ be coloured using the colours $1, 2, \ldots, r-1$ by giving an arc the colour i if it is in the orbital D_i. Now consider the set S of arcs (v, x) for all $x \in V$ different from v. Note that each of the $r - 1$ colours appears at least once on some arc in S because, if (a, b) is in D_i, then, since (Γ, V) is transitive, there is an $\alpha \in \Gamma$ such that $\alpha(a) = v$, and therefore $(v, \alpha(b))$ is in D_i and this has colour i.

Also, if x, y are in the same orbit under the action of Γ_v, then (v, x) and (v, y) must have the same colours. Conversely, if x, y are in different orbits under the action of Γ_v, then (v, x) and (v, y) must have different colours. That is, x and y are in the same suborbit if and only if (v, x) and (v, y) are in the same orbital.

Therefore, if the neighbours of v are grouped by orbits under the action of Γ_v, then the number of such groups will be equal to $r - 1$, the number of different colours appearing on arcs of S. In other words, the number of orbits under (Γ_v, V) excluding $\{v\}$ is equal to $r - 1$; that is, the number of suborbits is equal to the rank r. $\qquad\square$

Orbital graphs also provide a criterion for a permutation group to be primitive. This result is usually attributed to Higman and Sims.

Theorem 4.3 *Let (Γ, V) be a transitive permutation group. If some nontrivial orbital digraph of (Γ, V) is not strongly connected, then (Γ, V) is imprimitive. Also, if (Γ, V) is imprimitive, then some nontrivial orbital digraph is disconnected. Therefore (Γ, V) is primitive if and only if each orbital digraph is strongly connected.*

Proof Let $G = (V, D)$ be an orbital digraph. Let $\sim$ denote the equivalence relation on V defined by $u \sim v$ if and only if there are directed paths both from u to v and from v to u. If $u \sim v$, then, for any $\alpha \in \Gamma$, $\alpha(u) \sim \alpha(v)$, since α is an automorphism of G. Therefore the permutation group is compatible with this equivalence relation. Thus, if some nontrivial orbital digraph is not strongly connected, then (Γ, V) is imprimitive.

Conversely, let (Γ, V) be imprimitive and suppose that $\{V_1, \ldots, V_k\}$ is an imprimitivity block system. Let $u, v, w \in V_1$ with $u \neq v$. The orbital digraph G associated with (u, v) is nontrivial, and it is also disconnected, because no element of Γ can take the arc (u, v) into (w, w') or (w', w) with $w' \in V_i, i \neq 1$. $\qquad\square$

In order to illustrate this result consider again Example 1 for $n = 4$. Since the action of the cyclic group of order 4 is clearly imprimitive, according to this theorem we should have at least one orbital digraph that is not strongly connected (it is isomorphic to $2K_2$). However, it can happen, as in this case, that the other orbital digraphs are strongly connected.

Our final result in this section gives a criterion that ensures that at least one orbital is self-paired.

Theorem 4.4 *Let* (Γ, V) *be a transitive permutation group. If either r or* $|\Gamma|$ *is even, then there is a self-paired orbital.*

Proof If r is even, then there is an odd number of nontrivial orbitals. Since each orbital can be paired with its transpose, at least one must be self-paired. If $|\Gamma|$ is even, then the group has an element α of order 2. Hence there are $u, v \in V$ with $u \neq v$, $\alpha(u) = v$ and $\alpha(v) = u$. Thus $(u, v) \in D$ and $(v, u) \in D^T$ belong to the same orbital; therefore $D = D^T$. $\quad\square$

4.2 Rank 3 groups

Suppose that the transitive permutation group (Γ, V) has rank 3 and let its orbitals be D_0, D_1, D_2. If at least one of D_1 or D_2 is self-paired, then so is the other one. Henceforth, in this section, we shall consider this the case. Therefore both $G(D_1)$ and $G(D_2)$ are orbital graphs. Of course they are both regular (because Γ acts transitively on their vertex-sets) and they are complements of each other (because D_1 and D_2 are the only nontrivial orbitals). Moreover, if (Γ, V) is primitive, then both graphs are connected.

Theorem 4.5 *Let* (Γ, V) *be a transitive group of rank* 3 *and suppose D is a nontrivial self-paired orbital. Let* $G = G(D)$*. Then,*

  (i) *given any two adjacent vertices of G there is a constant number of vertices adjacent to both of them;*
  (ii) *given any two distinct nonadjacent vertices of G there is a constant number of vertices adjacent to both of them.*

Proof Suppose uv and $u'v'$ are edges of G. Since G is arc-transitive under the action of Γ, there is an $\alpha \in \Gamma$ such that $\alpha(u) = u'$ and $\alpha(v) = v'$. If w is adjacent to both u and v, then $\alpha(w)$ is also adjacent to both u' and v'. It is therefore clear that α induces a bijection between the set of vertices adjacent to both u and v and the set of vertices adjacent to both u' and v'. These two sets therefore have the same cardinality.

Now suppose u is not adjacent to v in G and neither are u' and v'. The complement $\overline{G}$ of G is the graph corresponding to the other orbital, and $uv, u'v'$ are both edges in $\overline{G}$. Therefore there exists $\overline{\alpha} \in \Gamma$ such that $\overline{\alpha}(u) = u'$ and

$\overline{a}(v) = v'$. Reasoning as earlier, one concludes that, in $\overline{G}$, the number of vertices not adjacent both to u and to v is the same as the number not adjacent both to to u' and to v'. Taking complements gives the result. $\qquad\square$

4.3 Strongly regular graphs

The result in the previous section motivates the following definition.

A graph G of order n is said to be *strongly regular* with parameters (n, k, λ, μ) (or, for short, we say that G is srg(n, k, λ, μ)) if

(i) G is regular of degree k;
(ii) for any pair v, w of adjacent vertices in G there is a constant number λ of vertices adjacent to both v and w;
(iii) for any pair v, w of distinct nonadjacent vertices of G there is a constant number of vertices μ adjacent to both v and w.

Although the definition of strongly regular graphs is motivated by the result in the previous section, it must be stated that a given strongly regular graph need not admit a group of automorphisms acting transitively and with rank 3 on its vertex-set. The smallest such example is the Shrikhande Graph, which has parameters $(16, 6, 2, 2)$. This graph can be constructed as the Cayley graph of the group $\mathbb{Z}_4 \times \mathbb{Z}_4$ with connecting set $S = \{\pm(1, 0), \pm(0, 1), \pm(1, 1)\}$. It is implemented in *Sage* and can be called up with the command

```
s := graphs.ShrikhandeGraph
```

Note, however, that any strongly regular graph is also a distance-regular graph with diameter 2.

We now first give some of the more elementary results on strongly regular graphs, most of whose proofs we leave as exercises.

Theorem 4.6 *Let G be* srg(n, k, λ, μ). *Then*

(i) $\overline{G}$, *the complement of G, is* srg$(n, \overline{k}, \overline{k} - k + \mu - 1, \overline{k} - k + \lambda + 1)$, *where* $\overline{k} = n - k - 1$.
(ii) G *is disconnected if and only if it is isomorphic to the disjoint union of r copies of K_{k+1} and this occurs if and only if $\mu = 0$.*
(iii) *If G is connected, then it has diameter 2.*
(iv) *If $\mu \neq 0$, then the parameters of G satisfy*

$$k(k - \lambda - 1) = (n - k - 1)\mu.$$

(v) *If $\mu = k$, then $\overline{\mu} = 0$, that is, the complement of G is disconnected.*

Proof We shall only prove (iv) and the rest are left as exercises. Fix a vertex v and count in two ways the number of pairs (x, y) such that x is adjacent to y and to v but y is not adjacent to v. Now, v has $n - k - 1$ non-neighbours y. Each such non-neighbour of v is adjacent to exactly μ neighbours x of v. Therefore the number of pairs (x, y) is $\mu(n - k - 1)$.

Now, v has k neighbours x. For each of these x there are λ common neighbours of x and v. But x has $k - 1$ neighbours in all (not counting v), so there are $k - \lambda - 1$ neighbours of x not adjacent to v. Therefore the number of pairs (x, y) is equal to $k(k - \lambda - 1)$. $\qquad\qquad\qquad\Box$

Examples

1) A graph that is a union of c copies of the complete graph K_{k+1} is $\mathrm{srg}(c(k + 1), k, k - 1, 0)$. Often, trivial examples such as this are excluded by assuming that a strongly regular graph and its complement are both connected. This is equivalent to assuming that

$$0 < \mu < k < n - 1.$$

2) The simplest nontrivial (in the above sense) strongly regular graph is the 5-cycle that has parameters $(5, 2, 0, 1)$.

3) The Petersen graph is $\mathrm{srg}(10, 3, 0, 1)$.

4) The *triangular graph* $T(r)$, $r \geq 4$ has as vertices the 2-element subsets of an r-set, and two vertices are adjacent if and only if they are not disjoint. $T(r)$ is $\mathrm{srg}(\binom{r}{2}, 2(r - 2), r - 2, 4)$. The Petersen graph is the complement of $T(5)$.

5) The *square lattice graph* $L_2(r)$, $r \geq 2$ has as a vertex-set $S \times S$, where S is an r-set, and two distinct vertices are adjacent if they have a common coordinate. $L_2(r)$ is $\mathrm{srg}(r^2, 2(r - 1), r - 2, 2)$.

4.4 The Integrality Condition

Clearly, not any set of numbers can be the parameters of a strongly regular graph. For example, they must satisfy the criterion of Theorem 4.6(iv), which is a result of simple counting arguments. We shall present here a stronger set of criteria obtained by a surprising and very pleasant use of linear algebra. These results were first obtained by Indian mathematicians working with R. C. Bose.

Lemma 4.7 *Let G be $\mathrm{srg}(n, k, \lambda, \mu)$ and let A be its adjacency matrix. Let I be the $n \times n$ identity matrix and J the $n \times n$ matrix, all of whose entries are equal* *to 1. Then*

$$A^2 + (\mu - \lambda)A + (\mu - k)I = \mu J.$$

Proof Each $(A^2)_{ij}$ equals the number of walks of length 2 between vertices v_i and v_j. Then

$$(A^2)_{ij} = \begin{cases} k & \text{if } i = j \\ \lambda & \text{if } v_i \text{ is adjacent to } v_j \\ \mu & \text{if } v_i \text{ is not adjacent to } v_j. \end{cases}$$

Therefore $A^2 = kI + \lambda A + \mu(J - I - A)$. That is, $A^2 = (\lambda - \mu)A - (k - \mu)I + \mu J$, as required. □

Corollary 4.8 *Let G be* $\mathrm{srg}(n, k, \lambda, \mu)$ *with* $\mu \neq 0$. *Then G has just three distinct eigenvalues: k, and* $s, t = \frac{1}{2}[\lambda - \mu \pm \sqrt{(\lambda - \mu)^2 + 4(k - \mu)}]$.

Proof Note first that, since G is regular of degree k, then $\mathbf{j} = (1, 1, \ldots, 1)^T$ is an eigenvector of the adjacency matrix A of G with eigenvalue k. Now suppose θ is another eigenvalue of A ($\theta \neq k$), with corresponding eigenvector $\mathbf{v}$. Thus, if $\mu \neq 0$ and applying the lemma to $\mathbf{v}$,

$$\frac{1}{\mu}[\theta^2 + (\mu - \lambda)\theta + (\mu - k)]\mathbf{v} = J\mathbf{v}.$$

But since eigenvectors of different eigenvalues of A are orthogonal and since all rows of J are equal to $\mathbf{j}$, $J\mathbf{v} = 0$. Therefore

$$\theta^2 + (\mu - \lambda)\theta + (\mu - k) = 0.$$

Solving this quadratic for θ gives the roots

$$\frac{1}{2}[\lambda - \mu \pm \sqrt{(\lambda - \mu)^2 + 4(k - \mu)}],$$

as required. □

Theorem 4.9 (Integrality Condition) *For a strongly regular graph G with parameters* (n, k, λ, μ), *the numbers:*

$$f, g = \frac{1}{2}\left[(n - 1) \pm \frac{(n - 1)(\mu - \lambda) - 2k}{\sqrt{(\mu - \lambda)^2 + 4(k - \mu)}}\right]$$

are nonnegative integers.

Proof Now let f and g be the multiplicities of the eigenvalues s and t, respectively. Then

$$f + g + 1 = n$$

(since the eigenvalue k has multiplicity 1). Also,

$$k \cdot 1 + f \cdot s + g \cdot t = 0$$

since the sum of the eigenvalues equals the trace of A, which is zero.

When these two equations are solved for f and g and the values of s and t substituted we obtain the values given in the statement, and since f and g are integers, the result follows.   □

We now present a partial converse of Corollary 4.8. The proof of this corollary again illustrates how results from linear algebra can give nontrivial results in graph theory.

Theorem 4.10 *Suppose G is not a complete graph and that it is connected and regular of degree k. Suppose also that G has just three distinct eigenvalues $k > s > t$. Then G is strongly regular.*

Proof We recall the following results from linear algebra about the adjacency matrix A of G:

- Since A is symmetric, it is similar to a diagonal matrix.
- All the distinct eigenvalues of a matrix are roots of its minimum polynomial.
- If a matrix is similar to a diagonal matrix, then its minimum polynomial factors into distinct linear factors.

These facts imply that the minimum polynomial of A is $(x - k)(x - s)(x - t) = (x - k)q(x)$. Therefore, $Aq(A) = kq(A)$. Hence each column of $q(A)$ is an eigenvector of A corresponding to the eigenvalue k. But k is a simple eigenvalue of A, since G is connected. Therefore k has only one eigenvector $\mathbf{j} = (1, 1, \ldots, 1)^T$ together with its multiples (see Exercise 2.14). Therefore, each column of $q(A)$ is a multiple $c_i \mathbf{j}$ of $\mathbf{j}$. Let $q(A) = J'$. But, since q is a polynomial of degree 2, this means that, for some constants a, b, c,

$$A^2 + aA + bI = cJ'.$$

Taking the ii–th entry gives

$$(A^2)_{ii} + b = c_i.$$

However, $(A^2)_{ii}$ is equal to the number of walks of length 2 from v_i to v_i, which is equal to k for all i, since G is k-regular. Therefore all the c_i are equal.

This means that, for some constants a, b, c,

$$A^2 + aA + bI = cJ.$$

But $(A^2)_{ij}$ equals the number of walks of length 2 joining vertices v_i and v_j. Therefore this equation implies that if v_i, v_j are adjacent, then they have $c - a$ common neighbours; if they are not adjacent, then they have c common neighbours; while if $v_i = v_j$, then the degree of v_i is $c - b$ (which therefore equals k). But this means that G is strongly regular. □

Note that the fact that G was regular was essential in this proof. In fact, the structure of nonregular graphs with three distinct eigenvalues has not been characterised and it is still a topic of active research (see [60, 199], for example).

4.5 Moore graphs

Recall that the diameter $d(G)$ of a graph G is defined as the maximum distance between any pair of vertices of G, while the girth $g(G)$ is the length of any shortest cycle in G. It is reasonable to expect that, given d and k, there is an upper bound on the number of vertices that a graph with diameter d and maximum degree k can have. This is confirmed by the following easy result.

Lemma 4.11 *If G is a connected graph with diameter d and maximum degree Δ, then*

$$n(G) \le n_0(d, \Delta) = 1 + \sum_{i=0}^{d-1} \Delta(\Delta - 1)^i.$$

Moreover, equality holds if and only if G is regular of degree Δ and has girth $2d + 1$.

Proof Let v_0 be a vertex of G which lies on the smallest cycle of G. Let us try to construct a graph with the largest number of vertices given that the graph has diameter d and maximum degree Δ.

So, v_0 has at most Δ neighbours. Each of the vertices has at most $\Delta - 1$ more vertices, and if we assume that no neighbours of v_0 are adjacent to each other and that these neighbours are distinct, then we have at this stage at most

$1 + \Delta + \Delta(\Delta - 1)$ vertices. Repeating this process we obtain that there are at most

$$1 + \sum_{i=0}^{r} \Delta(\Delta - 1)^i$$

vertices at the r-th stage. But since the diameter is d, this process cannot give us new vertices beyond $r = d - 1$. Therefore we obtain the upper bound, as required.

For equality to hold it is clear that every such graph should have degree Δ. Also, any vertex listed at the $d - 1$ stage has to have its remaining $\Delta - 1$ neighbours from amongst the other vertices listed at that stage. Since v_0 was chosen to lie on a minimum cycle of G, the girth of G must be $2d + 1$. $\qquad \square$

A regular graph of degree $k > 1$, girth $2d + 1$ and number of vertices equal to $n_0(d, k)$, as given by Lemma 4.11, is called a *Moore graph*.

We shall now focus our attention on Moore graphs of diameter 2, that is, girth 5.

Theorem 4.12 *If G is a Moore graph of diameter $d = 2$, then its degree k can only be equal to $2, 3, 7$ or 57.*

Proof A Moore graph of diameter 2 and degree k is $\mathrm{srg}(k^2 + 1, k, 0, 1)$. The Integrality Condition therefore becomes

$$f, g = \frac{1}{2} \left[k^2 \pm \frac{k^2 - 2k}{\sqrt{(4k - 3)}} \right]$$

are integers.

Therefore either $k^2 - 2k = 0$, giving $k = 2$, that is, G is the cycle on five vertices; or, $4k - 3 = r^2$, for some r. In the latter case, substituting for f gives

$$f = \frac{1}{2} \left[((r^2 + 3)/4)^2 + \frac{((r^2 + 3)/4)^2 - (r^2 + 3)/2}{r} \right].$$

Therefore

$$r^5 + r^4 + 6r^3 - 2r^2 + (9 - 32f)r - 15 = 0,$$

and r divides 15. Hence $r = 1, 3, 5$ or 15. Therefore $k = 3, 7$ or 57. $\qquad \square$

Note that the Petersen graph is a Moore graph with $k = 3$ and $d = 2$. It is also known that this is the only Moore graph with these parameters. Also, it is known that there is a unique Moore graph with $k = 7$ and $d = 2$, called the

Hoffman-Singleton Graph. It is, however, not known whether a Moore graph with $k = 57$ and $d = 2$ exists.

4.6 Exercises

4.1 In the action described earlier of S_n on the k-subsets of an n-set, when is it that there are orbitals that are not self-paired?

4.2 Show that the Petersen graph is the odd graph O_3.

4.3 Write down the intersection array of an $\mathrm{srg}(n, k, \lambda, \mu)$ considered as a distance-regular graph of diameter 2.

4.4 Prove the remaining parts of Theorem 4.6.

4.5 Prove that a strongly regular graph and its complement are both connected if and only if

$$0 < \mu < k < n - 1.$$

4.6 Let G be the graph obtained as follows. Let $A = \{1, 2, \ldots, 7\}$. Let the vertices of G be all the subsets of A of size 3, and let two vertices be adjacent if and only if the corresponding 3-subsets intersect in exactly one element. Show that G is strongly regular, and find its parameters.

4.7 Strongly regular graphs are sometimes classified into two types. Type 1 have the two multiplicities of the nonprincipal eigenvalues equal. That is, the numerator of the $\pm$ term in the Integrality Conditions is zero. Show that if G is strongly regular of Type 1, then G and $\overline{G}$ have the same parameters and $n = 4\mu + 1$. These graphs are called *conference graphs*. It is known that, for a conference graph, n is equal to the sum of two squares.

For Type 2 strongly regular graphs, that is, those for which the multiplicities of the nonprincipal eigenvalues are not equal, the term under the square root in the denominator of the $\pm$ term must be a perfect square r^2, and r must divide the numerator. Show that the quotient must be equal to $n - 1 \bmod 2$ and that the two eigenvalues are integers.

4.8 *The Friendship Theorem.* Let G be a graph on more than three vertices, and suppose that, for any two distinct vertices v and w in G, a unique vertex u is joined to both v and w. Following these steps, prove that G consists of a number of triangles with a common vertex.

   (a) Show that the degrees of any two nonadjacent vertices are equal. Therefore if G is not regular, then there exist adjacent vertices with different degrees.

   (b) Suppose that G is not regular. Deduce that it then indeed consists of a number of triangles with a common vertex.

   (c) Suppose G is regular. Show that it is a strongly regular graph with $\lambda = \mu = 1$.

   (d) In this case G has parameters $\mathrm{srg}(n, k, 1, 1)$. By considering the difference between the multiplicities f, g of the two nonprincipal eigenvalues, obtain that $k - 1$ divides k^2, therefore $k = 0$ or $k = 2$, that is G is either K_1 or K_3, which is not possible.

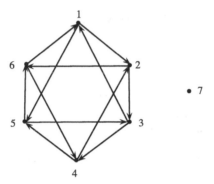

Figure 4.2. The graph G_0 for Exercise 4.10

4.9 A theorem of Sachs says that if G is a regular graph of degree k with n vertices and $m = \frac{1}{2}nk$ edges, and if the characteristic polynomial of G is $\phi(G; x)$, then the characteristic polynomial of its line-graph $L(G)$ is given by

$$\phi(L(G); x) = (x + 2)^{m-n} \phi(G; x + 2 - k).$$

Deduce that if G is strongly regular, then $L(G)$ cannot be strongly regular.

4.10 In more sophisticated uses of orbital graphs, the elements of V can themselves be combinatorial structures with an appropriate group of transformations acting on them. As an example we present the construction of the Klin Graph, a strongly regular graph on 210 vertices [39]. Let G_0 be the graph shown in Figure 4.2, and let V be the set of all nonidentical labellings of G_0 using the labels $1, \ldots, 7$, including the labelling shown in the figure.

The size of the automorphism group of G_0 is 24. This can easily be seen as follows. The vertices of G_0 can be partitioned into three subsets $\{1, 4\}$, $\{2, 5\}$ and $\{3, 6\}$, where the two vertices of the same subset have the same in- and out-neighbourhoods. Hence we have an elementary abelian group of order 8 with all permutations preserving the aforementioned partition. Then we have a cyclic group of order 3 that permutes cyclically the three subsets. The size of V (that is, the number of nonidentical labellings of G_0) therefore equals $7!/24 = 210$, by Exercise 4.2. Note that the large component of G_0 is isomorphic to the Cayley graph $\mathrm{Cay}(\Gamma, S)$ where Γ is the group generated by the permutation $\alpha = (1\ 2\ 3\ 4\ 5\ 6)$ and $S = \{a, a^4\}$ and the labelling of the outer 6-cycle corresponds to the permutation α. All the labellings of G_0 in V are obtained this way and are uniquely determined by a corresponding permutation $\alpha \in S_7$, which is a 6-cycle. We denote such a labelling by G_α, therefore G_0 is $G_{(1\ 2\ 3\ 4\ 5\ 6)}$.

Now let S_7 act in the natural way on the elements of V. The Klin Graph K is obtained by taking the union of seven orbitals of the action of S_7 on $V \times V$ for which the the following are representative arcs:

$$(G_0, G_{(1\ 2\ 4\ 5\ 6\ 7)}), (G_0, G_{(1\ 4\ 5\ 6\ 7\ 3)}), (G_0, G_{(1\ 2\ 5\ 4\ 6\ 7)}),$$
$$(G_0, G_{(1\ 4\ 2\ 5\ 6\ 7)}), (G_0, G_{(1\ 3\ 5\ 4\ 6\ 7)}), (G_0, G_{(1\ 4\ 3\ 5\ 6\ 2)}),$$
$$(G_0, G_{(1\ 3\ 5\ 4\ 6\ 2)}).$$

Use a computer package to construct the graph K and to verify that it is srg$(210, 99, 48, 45)$.

4.7 Notes and guide to references

The material presented here on orbital graphs is quite standard. More material on strongly regular graphs can be found in [254] and in the survey article [43]. For more on the link between strongly regular graphs and group theory see [24, 26, 49]. Reference [50] deals in much more detail with the link between strongly regular graphs, designs and codes.

The Friendship Theorem was first proved in [68]. The results from linear algebra on the minimum polynomial of a matrix can be found in [205].

An extensive treatment of minimal order regular graphs with given girth is given in [24]. This reference also contains a proof of Sach's Theorem [226] quoted in Exercise 4.9.

The book [49] gives a detailed description of how to construct the Hoffman-Singleton Graph using GAP. While it is not known whether a Moore graph of degree 57 and girth 5 exists, it is known that if such a graph does exist, then it cannot admit a transitive group of automorphisms with rank 3 [11], and, in fact, it cannot even be vertex-transitive [44, 49]. Some more recent results about the properties of a hypothetical Moore graph of degree 57 can be found in [156].

The Klin Graph was first described by Mikhail Klin at the fifteenth edition of the British Combinatorial Conference in 1995. A description of this graph explaining the computer experimentation which went behind the discovery of its construction as given in Exercise 4.10 can now be found in [125]. This paper gives various computer constructions of other graphs with interesting algebraic properties.

The proper context within which to study the construction given in Exercise 4.10 is *association schemes*. The paper [125] considers this and other constructions within the more general context of *coherent configurations*. The paper [72] is the main reference for information about COCO, which is a free computer package for working with coherent configurations and carrying out constructions such as the one given in Exercise 4.10. The paper [125] also explains the use of COCO, which can be downloaded either from Andreis Brouwer's website www.win.tue.nl/~aeb/ or from Dima Pasechnik's website https:// github.com/dimpase/coco.

Andreis Brouwer maintains a website [39] which lists several parameters of the strongly regular graphs on at most 1300 vertices, as well as of some larger ones and also of some interesting infinite families. Ted Spence's website [236] gives the adjacency matrices of the strongly regular graphs on at most 64 vertices, and Geoffrey Exoo's site [69] gives a list of feasible parameters for which the existence of a strongly regular graph has not yet been determined.

5

Graphical Regular Representations and Pseudosimilarity

We have seen that the full automorphism group of a Cayley graph $\text{Cay}(\Gamma, S)$ can be larger than Γ. If it happens that $\text{Aut}(\text{Cay}(\Gamma, S))$ is in fact equal to Γ, then, by Theorem 3.5, the automorphism group of $\text{Cay}(\Gamma, S)$ acts regularly on $V(\text{Cay}(\Gamma, S))$. Conversely, if G is a graph whose automorphism group acts regularly on $V(G)$, then G is a Cayley graph. (This follows from the second part of Theorem 3.5 and also from Sabidussi's Theorem because the stabiliser $\mathcal{H}$ in that theorem is trivial.)

A graph G whose automorphism group $\Gamma = \text{Aut}(G)$ acts regularly on $V(G)$ is called a *graphical regular representation (GRR)* of the group Γ. Therefore G is a GRR of its automorphism group Γ if it is vertex-transitive and the stabiliser Γ_v of any vertex is just the identity.

Note that in this case, by the Orbit-Stabiliser Theorem, $|V(G)| = |\Gamma|$. Also, given any two vertices u, v in G, there is one and only one automorphism α such that $\alpha(u) = v$. Also, $G - u$ has a trivial automorphism group for any vertex u.

5.1 Elementary results

Let us first consider the question of what conditions a graph must satisfy in order to be a GRR of its automorphism group. For convenience we shall state in the next theorem the main comments that we made in the introduction to this chapter, noting again that its proof follows from Theorem 3.5 or Sabidussi's Theorem.

Theorem 5.1 *Let G be a GRR of its automorphism group Γ. Then G is isomorphic to a Cayley graph $\text{Cay}(\Gamma, S)$ for some set S that generates the group Γ.*

Not any Cayley graph is, however, a GRR.

Lemma 5.2 *Let $G = \text{Cay}(\Gamma, S)$ and let ϕ be an automorphism of Γ such that $\phi(S) = S$. Then ϕ, as a permutation of $V(G) = \Gamma$, is an automorphism of G fixing the vertex 1.*

Therefore, if ϕ is nontrivial, then G is not a GRR.

Proof Since ϕ is a group automorphism, $\phi(1) = 1$. Also, ϕ is a graph automorphism because

$$\{\alpha, \beta\} \in E(G) \Leftrightarrow \alpha^{-1}\beta \in S$$
$$\Leftrightarrow \phi(\alpha^{-1}\beta) \in S$$
$$\Leftrightarrow \phi(\alpha)^{-1}\phi(\beta) \in S$$
$$\Leftrightarrow \{\phi(\alpha), \phi(\beta)\} \in E(G).$$

$\square$

These results therefore imply that, for G to be a GRR, it (i) must be isomorphic to a Cayley graph $\text{Cay}(\Gamma, S)$ and (ii) no nontrivial automorphism of Γ can fix the set S. In Exercise 5.2 at the end of this chapter you are asked to work an example that shows that these two necessary conditions are not sufficient to guarantee that G is a GRR.

5.2 Abelian groups

We now consider briefly the question of determining which groups have a GRR. This is a very difficult problem for which, however, a complete solution has been found through the efforts of many researchers. It is beyond the scope of this book to give the full solution, and we shall therefore state the complete theorem without proof. We can, however, give a complete treatment for the relatively simple case of abelian groups.

First let us look at a simple lemma from group theory whose proof was set as Exercise 1.13.

Lemma 5.3 *Let Γ be an abelian subgroup of S_Y and suppose that Γ acts transitively on Y. Then Γ acts regularly on Y.*

Theorem 5.4 *Let G be a graph and let $\mathcal{H}$ be an abelian subgroup of its automorphism group acting transitively on $V(G)$. Then $\mathcal{H}$ acts regularly on $V(G)$. Also, if $\mathcal{H}$ is the full automorphism group of G, then $\mathcal{H}$ is an elementary abelian 2-group.*

Proof The fact that $\mathcal{H}$ acts regularly on $V(G)$ follows from the previous lemma. Therefore G is isomorphic to a Cayley graph $\text{Cay}(\mathcal{H}, S)$ by Theorem 3.5.

Now suppose that $\mathcal{H}$ is the full automorphism group of G. Because $\mathcal{H}$ is abelian, the mapping $\alpha \mapsto \alpha^{-1}$ is an automorphism of $\mathcal{H}$. But this automorphism fixes S, and therefore it must be the identity, by Lemma 5.2. Therefore $\alpha = \alpha^{-1}$ for all elements $\alpha \in \mathcal{H}$; that is, $\mathcal{H}$ is an elementary abelian 2-group.

$\square$

We collect these facts into the following corollary.

Corollary 5.5 *Let the vertex-transitive graph G have a full automorphism group Γ. If Γ is abelian, then G is a GRR of Γ and Γ is an elementary abelian 2-group. Therefore, if $|\Gamma|$ is odd and G is a GRR of Γ, then Γ is non-abelian.*

While all abelian groups that have a GRR are elementary 2-groups, almost all such groups have GRRs. In Exercise 5.5 you are asked to prove the following theorem for $n \geq 5$.

Theorem 5.6 (Imrich) *Except for $n = 2, 3$ or 4, all elementary abelian 2-groups $\mathbb{Z}_2^n$ have GRRs.*

We finally state, without proof, the result that says which non-abelian groups have GRRs. Note that a *generalised dicyclic group* is a group Γ that is generated by an abelian subgroup $\mathcal{H}$ and an element $b \in \Gamma - \mathcal{H}$ such that $b^2 \in \mathcal{H}$ and $b^4 = b^{-1}hbh = 1$ for all $h \in \mathcal{H}$.

Theorem 5.7 *All finite non-abelian groups of odd order, with the exception of one group of order 27, have GRRs. Except for a finite number of known groups, all finite non-abelian groups of even order that are not generalised dicyclic groups have GRRs.*

In general, it is not easy to determine that a particular Cayley graph is a GRR of its automorphism group. A few relatively simple examples are given in the exercises.

5.3 Pseudosimilarity

Recall that two vertices u, v in G are pseudosimilar if they are removal-similar but not similar. An example of a graph with a pair of pseudosimilar vertices was given in Chapter 1.

In this chapter we shall see how some of the results that we have obtained so far can be used to construct graphs with pseudosimilar vertices.

5.4 Some basic results

Two sets of vertices A and B in a graph G are said to be *interchange-similar* if there is an automorphism α of G such that $\alpha(A) = B$ and $\alpha(B) = A$.

Lemma 5.8 *Let u, v be pseudosimilar vertices in a graph G, and let $A = N(u) \cap V(G - u - v)$, $B = N(v) \cap V(G - u - v)$. Then either A and B are similar but not interchange-similar in $G - u - v$ or else $G - v$ contains a vertex pseudosimilar to u.*

Proof Let $\alpha : G - v \to G - u$ be an isomorphism. If $\alpha(u) = v$, then restricting α to $V(G) - u - v$ gives an automorphism mapping A into B. Of course, A and B cannot be interchange-similar in $G - u - v$ because otherwise u and v would be similar in G.

We therefore assume that $\alpha(u) \neq v$. Let $w = \alpha^{-1}(v)$; therefore $w \neq u$. It now follows that u and w are removal-similar in $G - v$, for $(G - v) - w \simeq \alpha((G - v) - w) = (G - u) - v = (G - v) - u$.

Now suppose that u and w are similar in $G - v$. Let β be an automorphism of $G - v$ with $\beta(u) = w$. Then $\alpha\beta$ is an isomorphism from $G - v$ to $G - u$ with $\alpha\beta(u) = v$. We therefore again obtain that there is an automorphism of $G - u - v$ mapping A into B.

The only alternative left is that u and w are not similar in $G - v$, giving that $G - v$ contains a vertex pseudosimilar to u. $\qquad\square$

This easily gives, as a corollary, two well-known results about pseudosimilar vertices in a tree; the first of these was one of the earliest results on pseudosimilarity. We first need the following fact about similar vertices in a tree, whose proof is left as an exercise.

Theorem 5.9 (Prins) *Any two similar vertices in a tree are interchange-similar.*

We also recall that an *endvertex* is a vertex of degree equal to 1 and an *end-cutvertex* in a tree is a vertex having only one neighbour with degree greater than 1.

In the following theorem, part (a) is due to Harary and Palmer, while part (b) is due to Kirkpatrick, Klawe and Corneil.

Theorem 5.10 (Harary and Palmer; Kirkpatrick, Klawe and Corneil) *Let T be a tree. Then (a) any two removal-similar endvertices in T are similar and (b) any two removal-similar end-cutvertices in T are similar.*

Proof We shall prove (a) by induction on the number of vertices of the tree T; the proof of (b) is analogous. Suppose u, v are pseudosimilar endvertices of T. Let x, y be the neighbours of u, v, respectively. Since, by the induction hypothesis, u cannot be pseudosimilar to any endvertex in $T - v$, it follows from Theorem 5.8 that x and y are similar but not interchange-similar in $T - u - v$. But this contradicts Prins' Theorem. Therefore u and v cannot be pseudosimilar. $\qquad\square$

Harary and Palmer were the first to give a systematic way of constructing pairs of pseudosimilar vertices. Take any connected graph H, and let X and Y be two sets of vertices of H such that there is no automorphism α of H with $\alpha(X) = Y$ (for example, choose X and Y with $|X| \neq |Y|$). Then take three copies H_1, H_2, H_3 of H and form G by adding two new vertices u and v joining u to X in H_1 and to Y in H_2, and v to X in H_2 and to Y in H_3. Then u and v are pseudosimilar in G. The pair of pseudosimilar vertices shown in Figure 1.3 can be obtained this way, with $H = K_2$, $X = V(H)$ and Y any single vertex vertex of H.

However, the most general construction from which all pairs of pseudosimilar vertices can be obtained was found by Godsil and Kocay, who showed that such pairs can in fact always be obtained by destroying some cyclic symmetry in a graph. Take a graph H with vertices u and v and an automorphism α such that $\alpha^t(u) = v$ for some $t > 1$ and $\alpha^r(u) \neq v$ for $1 \leq r < t$. Then u and v are removal-similar in $G = H - \{\alpha(u), \ldots, \alpha^{t-1}(u)\}$; if, moreover, they also happen to be not similar, then we have a pair of pseudosimilar vertices. Godsil and Kocay showed that, in fact, every pair of pseudosimilar vertices can be obtained in this way. We give here this result without proof.

Theorem 5.11 (Godsil and Kocay) *Let u and v be a pair of pseudosimilar vertices in a graph G. Then G is an induced subgraph of some graph H such that H has an automorphism α with $\alpha(G - v) = G - u$ and $\alpha^t(u) = v$, and such that $V(H) - V(G) = \{x_1, \ldots, x_r\}$, where $x_i = \alpha^{t+i}(u)$ and $\alpha(x_r) = u$.*

Again, the pseudosimilar vertices shown in Figure 1.3 can be constructed using this manner of destroying a cyclic symmetry: here, H would be the graph shown in Figure 5.1 with $\alpha(u) = x$, $\alpha^2(u) = v$ and $\alpha^3(v) = u$.

But perhaps the most interesting problems in pseudosimilarity arise with graphs that have a large number of pseudosimilar vertices. This can happen in two ways. One way requires the graph to have several pairs of pseudosimilar vertices. Alternatively, one can ask for a graph with a large set of mutually

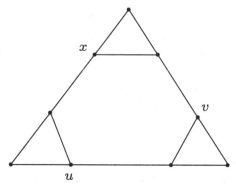

Figure 5.1. Illustration of Godsil and Kocay's theorem: creating a pair of pseudosimilar vertices by destroying a cyclic symmetry

pseudosimilar vertices (the vertices in a subset S of $V(G)$ are said to be *mutually pseudosimilar* if any two vertices in S are removal-similar but no two are similar).

These two problems will be considered in the next two sections.

5.5 Several pairs of pseudosimilar vertices

Kimble, Schwenk and Stockmeyer have shown that there exist graphs in which every vertex has a pseudosimilar mate! Their proof is based on the existence of GRRs.

Theorem 5.12 (Kimble, Schwenk and Stockmeyer) *There exist graphs G in which every vertex has a pseudosimilar mate; that is, for any $u \in V(G)$ there is a vertex $v \in V(G)$ such that u and v are pseudosimilar.*

Proof Let Γ be a group of odd order and let H be a GRR of Γ (as we have noted before, Γ must be non-abelian, and, by Theorem 5.7, such Γ and H do exist). We note that H is a regular graph and that the stabiliser of any vertex is just the identity element of Γ. Therefore, if r is any vertex of H, then $G = H - r$ has the identity automorphism group.

Now, let a be any vertex in G. There is an automorphism ϕ of H mapping r to a. The vertices $\phi^{-1}(r)$ and $a = \phi(r)$ are distinct, because otherwise ϕ would contain a cycle of length 2, which is impossible because Γ has odd order. Since ϕ^{-1} maps $\{a, r\}$ onto $\{r, \phi^{-1}(r)\}$, it follows that $G - a = H - r - a \simeq H - \phi^{-1}(r) - r = G - \phi^{-1}(r)$, that is, $a = \phi(r)$ and $\phi^{-1}(r)$ are removal-similar in G. But G has the identity automorphism group; therefore a and $\phi^{-1}(r)$ are pseudosimilar. □

The GRR constructed in Exercise 5.3 at the end of this chapter is an example of a graph H such that $G = H - r, r \in V(H)$, has all of its vertices paired by pseudosimilarity.

We observe that we have a stronger situation here than in Theorem 5.11 in the sense that adding a single vertex restores complete symmetry between the pairs of pseudosimilar vertices, making them all similar. So it is natural to ask if this always happens, that is, if graphs in which all vertices have a pseudosimilar mate are always obtained using the method of Theorem 5.12.

5.6 Several pairs of pseudosimilar edges

Note that pseudosimilar edges can be defined analogously to pseudosimilar vertices. We can adopt the same strategy as the Kimble, Schwenk and Stockmeyer construction to construct graphs in which every edge has a pseudosimilar mate. But to do this we have to construct a graph whose automorphism group has odd order and acts regularly on the edges. Therefore, by going to the line-graph, we need a Cayley graph Cay(Γ, S) whose full automorphism group is Γ and which is also a line-graph. For this we cannot just use the existence of GRRs for non-abelian groups because Cayley graphs which are line-graphs have a connecting set of a very particular type (see Exercise 3.14) and such a connecting set is not guaranteed to give a GRR in general.

Exercise 5.15 shows an infinite family of graphs which Alspach and Xu have shown to possess an odd order automorphism group acting regularly on the edges and therefore deleting an edge gives a graph all of whose edges are paired by pseudosimilarity. These graphs are both vertex- and edge-transitive. Here, we shall use the computer package GAP to construct such a graph which is, however, edge- but not vertex-transitive and whose automorphism group acts regularly on the edges. The graph we construct is a double coset graph.

First consider the group

$$\Gamma = \langle a, b, c | a^5 = b^3 = c^3 1, ba = abc, ca = ac^2, cb = bc^{25} \rangle.$$

This group has order $3 \times 5 \times 31 = 465$. Our graph will be the double coset graph Cos($\Gamma, \mathcal{H}, \mathcal{K}$) where $\mathcal{H} = \langle a \rangle$ and $\mathcal{K} = \langle b \rangle$. Clearly this double coset graph is not vertex-transitive since the vertices in the two colour classes have degrees 3 and 5. By Theorem 3.9 we know that it is edge-transitive. We shall use GAP to show that its automorphism group has order 465 therefore it acts regularly on the edges.

So, after loading GAP and the package GRAPE, we construct Γ, $\mathcal{H}$, $\mathcal{K}$ and the cosets on which Γ will act as follows.

```
gap> f:=FreeGroup(3);;
gap> rels:=[f.1^5,f.2^3,f.3^31,
       f.2*f.1*(f.1*f.2*f.3)^-1,
       f.3*f.1*(f.1*f.3^2)^-1,
       f.3*f.2*(f.2*f.3^25)^-1];;
gap> gamma:=f/rels;;

gap> h:=Subgroup(gamma,[gamma.1]);;
gap> k:=Subgroup(gamma,[gamma.2]);;

gap> hcoset:=RightCoset(h,gamma.1);;
gap> kcoset:=RightCoset(k,gamma.1);;
```

The double coset graph is then constructed using GRAPE's 'Graph' function as we did in Chapter 3.

```
gap> grph:= Graph(gamma,[hcoset,kcoset], OnRight,
       function(x,y) return x<>y and Intersection(x,y)<>[];
       end);
```

The following then confirms that the size of the automorphism group of grph is equal to the number of edges.

```
gap> grp=AutGroupGraph(grph);;
gap> Order(grp);
465
```

Therefore by deleting an edge from the coset graph we obtain a graph on 464 edges all of which are paired by pseudosimilarity. The smallest graph constructed by the method of Exercise 5.15 also has 464 edges. Therefore the natural question here is: are there smaller graphs in which every edge has a pseudosimilar mate?

5.7 Large sets of mutually pseudosimilar vertices

A graph G cannot have all of its vertices mutually pseudosimilar (see Exercise 5.9). Therefore one question that arises is to determine the largest size that a set of mutually pseudosimilar vertices in a graph on n vertices can have. This seems a very difficult problem in general. The smallest asymmetric graph, on eight vertices, with a pair of pseudosimilar vertices is shown in Figure 5.2,

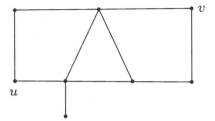

Figure 5.2. The smallest asymmetric graph with a pair u, v of pseudosimilar vertices

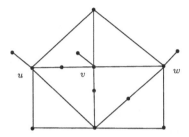

Figure 5.3. A smallest graph with three mutually pseudosimilar vertices u, v, w

whereas Figure 5.3 shows Exoo's Graph, which is a smallest graph (on 13 vertices) containing three mutually pseudosimilar vertices. More examples can be found on Exoo's website [69].

We describe here a method that constructs graphs with large sets of mutually pseudosimilar endvertices. This method also uses results that we have described in earlier chapters. In the exercises we describe extensions of this construction and also another method due to Kimble, Schwenk and Stockmeyer.

Suppose Γ is a group of permutations acting on some set X such that, for some $R \subset X$, the following two conditions hold:

A the setwise stabiliser $\Gamma_{\{R\}}$ of R is the identity;

B for any two $(|R| - 1)$-subsets A, B of R, there is a permutation α in Γ such that $\alpha(A) = B$.

Then, by Bouwer's Theorem, one can construct a graph G having $X \subseteq V(G)$ such that the action of $\mathrm{Aut}(G)$ on X is equivalent to the action of Γ on X. Note that it is easy to make sure that the graph G constructed in the proof of Bouwer's Theorem has minimum degree at least 2.

Therefore, attaching one endvertex to each vertex of $R \subset V(G)$ gives a graph G', all of whose endvertices are mutually pseudosimilar. The graphs G and G' can now be used as the basis of a construction of an infinite family of graphs with an increasing number of mutually pseudosimilar endvertices. Let $G_1 = G'$ and let H_1 be G_1 less one of its endvertices. Having constructed G_t, let H_t be G_t less one of its *pseudosimilar* endvertices. The graph G_{t+1} is obtained by taking G and attaching a copy of G_t to each vertex in $R \subset V(G)$ and a copy of H_t to each of the other vertices in $X - R \subset V(G)$. (By 'attaching' a copy of G_t (or H_t) to a vertex v of G we mean here joining v to every vertex in a union of orbits under the action of $\mathrm{Aut}(G_t)$ (or $\mathrm{Aut}(H_t)$) on the vertices of G_t (or H_t) that are *not* endvertices.)

What proportion of vertices in G_t are pseudosimilar? This graph has $|R|^t$ mutually pseudosimilar endvertices and $O(|X|^t)$ vertices in all. Therefore, if $k = |R|^t$ is the number of pseudosimilar endvertices, then the order of G_t is $O(k^{\log |X| / \log |R|})$.

We summarise the result of this construction in the following theorem.

Theorem 5.13 *The aforementioned construction starting with a permutation group Γ satisfying conditions* **A** *and* **B** *produces a family of graphs G_t, $t \geq 1$, with k mutually pseudosimilar endvertices and order*

$$O(k^{\log |X| / \log |R|}).$$

To carry out this construction one needs a permutation group having the conditions stated. Some examples are provided in the exercises.

5.8 Exercises

5.1 Let G be a Cayley graph $\mathrm{Cay}(\Gamma, S)$ and let $\mathrm{Aut}(\Gamma, S)$ be the set of all group automorphisms of Γ that fix S setwise. Prove that $\mathrm{Aut}(\Gamma, S)$, considered as a set of permutations of the vertices of G, is a subgroup of $\mathrm{Aut}(G)$. (By Lemma 5.2, $\mathrm{Aut}(\Gamma, S)$ is a subset of $\mathrm{Aut}(G)_1$, the stabiliser of 1 in $\mathrm{Aut}(G)$.)

Show that $L(\Gamma)$, the left regular representation of Γ on itself, is a normal subgroup of $\mathrm{Aut}(G)$ if and only if $\mathrm{Aut}(\Gamma, S) = \mathrm{Aut}(G)_1$.

5.2 Let Γ be the dihedral group of order 12 defined by

$$\Gamma = \langle a, \beta \,|\, a^2 = \beta^6 = 1, \beta a = a\beta^5 \rangle.$$

In Exercise 3.11(b) you were effectively asked to show that $\mathrm{Cay}(\Gamma, S)$ is not a GRR, where $S = \{a, a\beta, a\beta^3\}$. Show, however, that there is no nontrivial automorphism of Γ that fixes the set S and so deduce that the converse of Lemma 5.2 is false.

5.3 Let Γ be the metacyclic group of order 21 defined by

$$\Gamma = \langle a, \beta \,|\, a^3 = \beta^7 = 1, \beta a = a\beta^2 \rangle.$$

Let G be the Cayley graph $\mathrm{Cay}(\Gamma, S)$, where

$$S = \{a, a^{-1}, \beta, \beta^{-1}, a\beta, a^{-1}\beta^3, a\beta^5, a^{-1}\beta\}.$$

The aim of this exercise is to show that G is a GRR.

All we have to show in order to prove this is that the only automorphism of G that fixes 1 is the identity. Let ϕ be an automorphism of G that fixes 1 and let H be the subgraph of G induced by the neighbours of 1. Therefore ϕ fixes H and is an automorphism of H. Considering this subgraph, show that β and β^{-1} must also be fixed by ϕ. Now similarly consider the subgraph induced by the neighbours of any vertex γ and deduce that, if ϕ fixes γ, then it also fixes $\gamma\beta$ and $\gamma\beta^{-1}$. From the graph H deduce also that ϕ either fixes α and α^{-1} or it reverses them. Finally, deduce that ϕ fixes all vertices of G, as required.

5.4 Let $G = \text{Cay}(\Gamma, S)$ be a Cayley graph, and suppose that the automorphism group of the subgraph of G induced by the vertices in S is the trivial automorphism group. Show that G is a GRR of Γ.

5.5 (Imrich [109]) The aim of this exercise is to show that, for $n \geq 5$, the elementary abelian 2-group $\Gamma = \mathbb{Z}_2^n$ has a GRR. Let Γ be generated by the distinct elements $a_i, 1 \leq i \leq n$, and let S be the set

$$\{a_i, a_k a_{k+1}, a_1 a_2 a_{n-2} a_{n-1}, a_1 a_2 a_{n-1} a_n : 1 \leq i \leq n, 1 \leq k < n\}.$$

Consider the Cayley graph $G = \text{Cay}(\Gamma, S)$. Suppose ϕ is an automorphism of G that fixes 1. Let H be the subgraph of G induced by the vertices adjacent to 1. Therefore $V(H)$ is invariant under ϕ. Show that H has the trivial automorphism group, and therefore ϕ fixes all vertices of G.

5.6 Let $G = \text{Cay}(\Gamma, S)$ be the Cayley graph of the dihedral group

$$D_7 = \langle \alpha, \beta | \alpha^2 = \beta^7 = 1, \beta\alpha = \alpha\beta^6 \rangle$$

and let

$$S = \{\alpha, \alpha\beta, \alpha\beta^3, \beta, \beta^6\}.$$

Show that G is a GRR of D_7. [You might care to use one of the computer packages mentioned in Chapter 2 to help out with this.]

5.7 Show that an asymmetric graph need not have property A_1.

5.8 Let the graph $G = \text{Cay}(\Gamma, S)$ be a GRR of the group Γ. Let T be another subset of Γ with $T^{-1} = T$ and let $H = \text{Cay}(\Gamma, T)$. Suppose σ is an isomorphism from the graph G to the graph H such that $\sigma(1) = 1$. Show that σ is an automorphism of the group Γ such that $\sigma(S) = T$.

5.9 Show that a regular graph cannot have pseudosimilar vertices. Deduce that not all vertices can be mutually pseudosimilar. Show also that, if G has n vertices, then it also cannot have $n - 1$ mutually pseudosimilar vertices.

5.10 Let u and v be a pair of pseudosimilar vertices in a graph G, and let H be a graph with a distinguished vertex a. Denote by $G_u H$ and $G_v H$ the graphs obtained by identifying the vertex a of H with vertices u and v, respectively. Kocay showed, as follows, that it is possible for $G_u H$ and $G_v H$ to be isomorphic:

Let A_4 denote the alternating group acting on $X = \{u, v, w, x\}$. Use Bouwer's Theorem to construct the corresponding graph K with $X \subseteq V(K)$ and such that the action of $\text{Aut}(K)$ on X is equivalent to that of A_4. Let $G = K_x H - w$. Show that u and v are pseudosimilar in G and that $G_u H \simeq G_v H$.

5.11 Krishnamoorthy and Parthasarathy were the first to use the construction described in Section 5.7 to obtain graphs with many mutually pseudosimilar endvertices.

T_4:

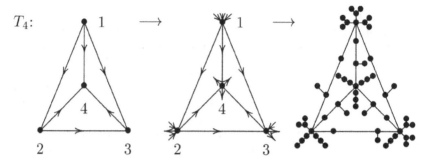

Figure 5.4. Four mutually pseudosimilar vertices from the transitive tournament on four vertices

Their particular group Γ was the cyclic group acting on $X = \{u, v, x\}$ generated by the permutation $(u\ v\ x)$. With $R = \{u, v\}$ this clearly satisfies conditions **A** and **B**. Their starting G graph can be taken as the one shown in Figure 5.1.

Using this graph as the basis of the construction described earlier, obtain the graph G_3 having 2^3 mutually pseudosimilar endvertices.

5.12 Kimble, Schwenk and Stockmeyer described another way to construct graphs with many mutually pseudosimilar vertices. The *transitive tournament* on k vertices, T_k, is the tournament with vertex-set $\{1, \ldots, k\}$ and in which vertex i dominates vertex j if and only if $i < j$. Clearly, the vertices of T_k are all mutually pseudosimilar. The problem, therefore, is how to transform T_k into an undirected graph while preserving the pseudosimilarity of its vertices. The arcs can be replaced by a gadget as usual, but a 'correction' would have to be made at each of the vertices of T_k.

Figure 5.4 illustrates this starting with the transitive tournament T_4 to obtain a graph with four mutually pseudosimilar vertices. Show that this method constructs a sequence of graphs G_k with k mutually pseudosimilar vertices and order $O(k^2)$.

5.13 Show that the only tournaments in which all vertices are mutually pseudosimilar are the transitive tournaments.

5.14 Let $X = F^{r-1}$, the vector space of dimension $r - 1$ over the finite field F, and let Γ be the group of all linear transformations on X. Let $B = \{e_1, \ldots, e_{r-1}\}$ be a basis of $X, f = \sum a_i e_i$ an element of X, and $R = B \cup \{f\}$. Suppose the following conditions on the a_i hold:

(a) $a_i \neq 0, 1 \leq i \leq r - 1$;
(b) $a_i \neq a_j, i \neq j$;
(c) $a_i a_j \neq 1, 1 \geq i, j \geq r - 1$;
(d) $a_i + a_j a_k \neq 0, 1 \leq i, j, k \leq r - 1$.

Show that Γ and R have properties **A** and **B** described in the last section of this chapter. Show also that if $q = |F|$ is greater than some quadratic polynomial in r, then it is always possible to choose the a_i satisfying conditions (i)–(iv).

5.15 Let p be a prime number congruent to 1 mod 3. Let $d > 1$ be a divisor of $(p - 1)/3$ such that 3 does not divide d. Let $T = \langle t \rangle$ be the subgroup of $\mathbb{Z}_p^*$ ($= \mathbb{Z}_p - \{0\}$)

of order d, and let $r \in Z_p^* - T$ have order 3 in Z_p^*. Let $\mathcal{H}_{3p}$ be the group

$$\mathcal{H}_{3p} = \langle \alpha, \beta | \alpha^p = \beta^3 = 1, \alpha^\beta = \alpha^r \rangle$$

and let

$$S = \{\beta\alpha^i : i \in T\} \cup \{\beta^2\alpha^{r^2 i} : i \in T\}.$$

Alspach and Xu have shown that the full automorphism group of the Cayley graph of $\mathcal{H}_{3p}$ with respect to S acts regularly on the edges of the graph.

Deduce that there is an infinite family of graphs such that, in each, every edge has a pseudosimilar mate. How many edges does the smallest graph of this family have?

5.9 Notes and guide to references

The efforts of several authors contributed to proving Theorem 5.7, which solved the problem of the existence of GRRs. The reader is especially referred to [86, 105, 111, 112, 115, 116, 208, 209]. The book [57] contains a census of all cubic GRRs of order at most 120. The last part of Exercise 5.1 (on the normality of $L(\Gamma)$) is taken from [263]. Exercise 5.3 is taken from [104] and Exercise 5.5 is taken from [109].

While, as we have seen, the determination of those groups with a GRR has been solved, the question whether a group has a GRR with some prescribed restriction like, say, being cubic, is still open. Questions of this type are considered in [71, 87, 148, 264].

Exercise 5.8 is taken from [185], and it is related to Ádám's Conjecture, which is discussed briefly in the notes at the end of Chapter 7.

Pseudosimilarity arose as a counterexample to a purported proof of the Reconstruction Conjecture (which will be discussed in a later chapter). The first serious study of pseudosimilarity occurred in [100, 101]. More on pseudosimilarity in trees can be found in [122]. Theorem 5.11 is from [88]. Theorem 5.12 is from [121]. Exercises 5.10 and 5.11 are from [129] and [135], respectively. The construction in Exercise 5.14 is from [137], and it is based on a construction of P. J. Cameron. In [9] it is shown that the automorphism groups of the graphs in Exercise 5.15 act regularly on their edges, and this was then used in [146] to give graphs all of whose edges are paired by pseudosimilarity. Two surveys on pseudosimilarity in graphs are given in [138, 139].

The analogue of Exercise 5.9 for locally finite infinite graphs is highly nontrivial, and its proof can be found in [245].

6

Products of Graphs

This chapter will deal with operations on graphs; these operations will be generically named 'products'. The word is already in use for several known operations, like the direct product and the cartesian product. Our approach will try to be unifying; that is, we will offer a mathematical definition of what a 'graph product' is. All we require for such an operation is (1) to have the vertex-set as the cartesian product of the vertex-sets of the factors involved in the operation, and (2) to have adjacencies within the resulting product depending solely on the adjacencies inside the factors, according to certain given rules.

Our starting point will be the direct product (also often referred to in the literature as the categorical product), whose definition may be considered the simplest, and upon which we shall obtain all the other products as graph-theoretic unions of suitable direct products.

Most of the content of the present chapter is quite well known, but the results concerning the various graph products that we shall present will often have to be rephrased in order to be consistent with the scheme presented earlier.

We shall especially focus on how the connectedness of the product (as well as other matters, such as those concerning paths) is related to that of the factors. We shall also stress the properties of products as binary operations. Incidentally, let us recall that we consider finite graphs only. Accordingly, some refinements that make sense in the infinite case, such as the distinction between 'weak' and 'strong' products, will not be our concern. Also, the associative property shall be considered only from this finitary viewpoint.

Here is how this chapter is organised. The general definitions that are typical of our approach will be recalled in Section 6.1. The remaining sections will deal with properties of remarkable special cases, namely the direct product (Section 6.2); the cartesian, strong and lexicographic product (Section 6.3); the modular product; and the 'Knight' product (Section 6.4).

6.1 General products of graphs

We start with the definition of the direct product. Let G and H be two graphs. The *direct product* $G \times H$ is defined as the graph whose vertex-set is $V(G) \times V(H)$ and such that (u_1, v_1) and (u_2, v_2) are adjacent if and only if u_1, u_2 are adjacent in G and v_1, v_2 are adjacent in H. Likewise, if G and H are digraphs, their direct product still has $V(G) \times V(H)$ as a vertex-set, and there is an arc from (u_1, v_1) to (u_2, v_2) if and only if there is, in G, an arc from u_1 to u_2 and, in H, an arc from v_1 to v_2.

Now, let $\mathcal{G}$ be a class of graphs (directed or undirected). A *twisting graph function* (TGF) is a map $f : \mathcal{G} \to \mathcal{G}$ such that $f(G)$ has the same vertex-set as G, and f takes isomorphic graphs into isomorphic graphs.

There are several possible examples of TGFs. We will list here three that have some relevance in the forthcoming discussions.

id: the identity map, $id(G) = G$;
cp: the map taking a loop-free graph G to its complement, $\overline{G}$;
loop: the map taking each G into a graph consisting only of loops, G°.

Further examples of a TGFs can be obtained as follows. Let k be a fixed nonnegative integer. For every graph G define G^k as the graph on $V(G)$ where two vertices a, b are adjacent if and only if $d(a, b)$, the distance between a and b, equals k. When $k = 0$, we get the graph whose edge-set consists only of all the loops: this is the TGF that we introduced earlier with the name *loop*. When k is larger than the diameter of G (or the maximum of the diameters of the components of G when G is disconnected), we obtain the null graph. If $\mathcal{G}$ is the class of all graphs, then the function on $\mathcal{G}$ taking each G to G^k is a TGF.

A *general graph product* (GGP), or simply a *product*, is a map

$$p : \mathcal{G} \times \mathcal{G} \to \mathcal{G}$$

such that the vertex-set of $p(G, H)$ is $V(G) \times V(H)$, and the arc set of $p(G, H)$ takes the form

$$\bigcup_{i=1}^{r} f_i(G) \times f_i'(H),$$

where all the f_i and f_i' are TGFs and r is a positive integer called the *rank* of the product.

A survey on known products has shown that, to our best knowledge, most of the graph products whose carrier is the cartesian product of the vertex-sets fit within this scheme provided that the product depends only on the factors. For example:

The direct product: $k = 1, f_1 = f_1' = id$;
The cartesian product: $k = 2, f_1 = f_2' = id, f_2 = f_1' = loop$;
The strong product: $k = 3, f_1 = f_2' = id, f_2 = f_1' = loop, f_3 = f_3' = id$.

The same may be applied to further operations, like the cartesian sum $G \oplus H$ (introduced by Ore [212]), that we can define equivalently as $G \oplus H = (\overline{G} \times H) \cup (G \times \overline{H})$, where $G \times H$ denotes the direct product.

Usually, the class $\mathcal{G}$ will simply be that consisting of all graphs or of all digraphs. However, we may consider other cases where the TGFs make sense only for suitable restrictions of either basic class.

In view of our definitions, the following statement is easy to prove. We leave the details to the reader.

Theorem 6.1 *Let $\mathcal{G}$ be a class of graphs or digraphs and let $*$ be a GGP defined on $\mathcal{G}$. Assume that all TGFs f involved in $*$ are such that $\mathrm{Aut}(f(G))$ contains $\mathrm{Aut}(G)$ as a subgroup. Then, for every $G, H \in \mathcal{G}$, the group $\mathrm{Aut}(G * H)$ contains a subgroup isomorphic to $\mathrm{Aut}(G) \times \mathrm{Aut}(H)$.*

Let us illustrate the use of GGPs by showing how the Petersen graph can be obtained as such a GGP of rank 3. Let us premise that the idea of a 'complement' of a graph can be considered from three different viewpoints. The usual one (that gives rise to the definition of *cp*) is standard with simple, undirected graphs: loops do not even exist, so they are duly ignored. Another one is to consider loops as well as arcs and consider the complement of G to be the graph whose arc-set is the complement of the arc-set of G: this gives a TGF *cpl* which takes G to its complement in this sense. Therefore an edge $\{u, v\}$ in G is removed completely in $cpl(G)$, and arc (u, v) without the inverse arc (v, u) in G becomes the arc (v, u) in $cpl(G)$, and there is a loop at a vertex v in $cpl(G)$ if and only if there is no loop in G.

Taking this idea further, we can ignore arcs and edges and only deal with loops, so the 'complement' of G in this third sense is the graph with the same 'proper' arcs as G, but whose set of loops is the complement of the set of loops of G. We shall call f^c the corresponding TGF. This particular map will help us in the following construction. Define a new TGF f such that $f(G)$ is a graph whose arc set consists of a single loop. (Of course, there are many ways to choose the vertex of G at which there is the loop, but in the sequel any such choice will give the same result.) Let $f_1 = id, f_2 = cp, f_3 = loop, f_1' = f, f_2' = f^c$ and $f_3' = id$. With this choice of TGFs, the corresponding product $*$ of rank 3 gives, when $G = C_5$, the cycle on five vertices, and $H = K_2$, a graph $G * H$ that is isomorphic with the Petersen graph. This construction is illustrated in

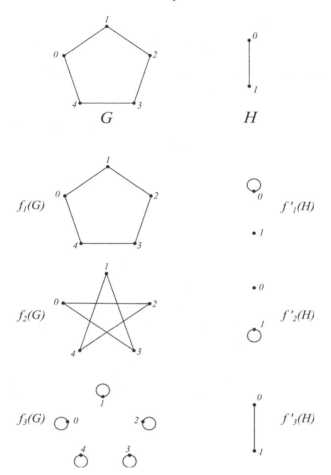

Figure 6.1. The graphs and TGFs involved in the construction of the Petersen graph as a GGP

Figure 6.1. Note that both G and H are Cayley graphs while their product is not. So, a GGP of Cayley graphs need not be a Cayley graph.

6.2 Direct product

We shall first study some of the simpler properties of our basic graph product. Due to the nature of the involved operations, the product of two connected graphs is not necessarily connected (see, for example, Theorems 6.2 and 6.10, among others). Vice versa, a product of disconnected graphs may be connected (see Exercise 6.2).

A product $*$ will be called *connectedness preserving* when the connectedness of the result relies on that of its factors, namely: $G * H$ is connected if and only if both G and H are connected. As we shall see later, the cartesian product enjoys this property. This is not the case for the direct product, as the following result shows. Recall that $G \times H$ denotes the direct product of G and H.

Theorem 6.2 *Let G and H be two connected and bipartite graphs. Then $G \times H$ has exactly two connected components.*

Proof Let $\alpha : V(G) \to \mathbb{Z}_2$ and $\beta : V(H) \to \mathbb{Z}_2$ be 2-colorings of G and H, respectively. Define

$$\gamma : V(G) \times V(H) \to \mathbb{Z}_2$$

$$\gamma : (x, y) \mapsto \alpha(x) + \beta(y).$$

That (x, y) and (a, b) are adjacent implies that $\gamma(x, y) = \gamma(a, b)$. This implies that if (x, b) and (a, y) are in the same connected component, then $\gamma(x, y) = \gamma(a, b)$. At this point we are left with two sets $\gamma^{-1}(0)$ and $\gamma^{-1}(1)$ that have no connection with each other.

We have now to prove that $\gamma^{-1}(0)$ and $\gamma^{-1}(1)$ are both connected. Consider $\gamma^{-1}(0)$. Let us fix $(x, y), (a, b) \in \gamma^{-1}(0)$, and let $x, x', \ldots, a$ and $y, t', \ldots, b$ be walks in the respective graphs. Their lengths have the same parity. Since $(x, y) \sim (x', y')$, we can replace (x, y) by (x', y') if they are odd. Therefore, we may assume that the lengths are both even. Now

$$(x, y), (x_2, y'), (x_3, y), \ldots, (a, y)$$

is a path. Similarly,

$$(a, y), (a', y_1), (a, y_2), \ldots, (a, b)$$

is a path. Joining the two given paths, we obtain a path from (x, y) to (a, b). This proves that $\gamma^{-1}(0)$ is connected. The same argument applies to $\gamma^{-1}(1)$. $\qquad\square$

Theorem 6.3 *If G and H are connected and G is not bipartite, then $G \times H$ is connected.*

Proof Let G' and H' be spanning trees of G and H, respectively. Take 2-colourings α of G' and β of H' and define γ as in the previous theorem. As earlier we can prove that $\gamma^{-1}(0)$ and $\gamma^{-1}(1)$ are both connected.

Let $(x, y) \in \gamma^{-1}(0)$ and $y' \sim y$, such that x is on an odd cycle. Thus $(x, y') \in \gamma^{-1}(1)$. Pick an odd length cycle $x = x_1, \ldots, x_m = x$ and consider

$$(x, y), (x_2, y'), (x_3, y), \ldots, (x_m, y') = (x, y').$$

Then $(x, y) \sim (x, y')$, and this proves that $\gamma^{-1}(0)$ and $\gamma^{-1}(1)$ are connected to each other. $\qquad\Box$

6.3 Cartesian product

We shall introduce here three remarkable graph products.

Let G and H be digraphs. The *cartesian product* of G and H is $G \square H = (G \times H^\circ) \cup (G^\circ \times H)$. The *strong product* of G and H is $G \boxtimes H = (G \times H) \cup (G \square H)$. The *lexicographic product* of G and H is $G \propto H = (G^\circ \times H) \cup (G \times H)$.

It is known that the cartesian product shares the associative property with these and many other products of graphs (see [113] for a study on the associativity of such operations). On the other hand, this product is prominent because of other peculiarities.

Apart from significant special cases, such as the Hamming graphs (cartesian products of complete graphs) or the hypercubes (that arise when all factors of the product are K_2), it has been proved by Sabidussi that the decomposition of a graph into a cartesian product is unique in the sense of the following statement. Here, a (finite) graph is called *prime* when it is not isomorphic to a cartesian product of smaller graphs.

Theorem 6.4 (Sabidussi) *Every connected graph can be written as a cartesian product of prime graphs. Such a decomposition is unique up to a reordering of factors.*

Other products can hardly satisfy such a strict property. For example, it is fairly easy to produce counterexamples with the direct product, even having just K_2 as one of the factors (see Exercise 6.5 at the end of this chapter).

While some of these operations are considered in [52] for their applications to the design of 'multistage' connection networks, a remarkable property of the cartesian product is recalled there. Namely, if κ_1, κ_2, κ and λ_1, λ_2, λ denote the vertex and the edge connectivity of G_1, G_2 and their cartesian product G, respectively, we have $\kappa_1 + \kappa_2 = \kappa$ and $\lambda_1 + \lambda_2 = \lambda$. The property extends to the case of several factors, due to the associativity of the cartesian product.

Let G be a graph and let $u, v \in V(G)$. The number of shortest paths between u and v will be denoted by $\gamma(u, v)$. The set-theoretical union of all shortest paths between vertices u and v is denoted by $I(u, v)$ and is called the *interval* between u and v.

Lemma 6.5 *Let $G = G_1 \square G_2$, with G_1 and G_2 both connected graphs. Then G is connected.*

Furthermore, if $x = (x_1, y_1)$, $y = (x_2, y_2)$ are two vertices of G and $d(x_1, x_2) = k_1$ and $d(y_1, y_2) = k_2$, then $d(x, y) = d(x_1, y_1) + d(x_2, y_2)$ and

$$\gamma\left((x_1, y_1), (x_2, y_2)\right) = \binom{k_1 + k_2}{k_1} \gamma_1(x_1, x_2) \gamma_2(y_1, y_2).$$

Proof The first part of the proof is left to the reader as an exercise (see Exercise 6.4).

Now let $x = (x_1, y_1)$ and $y = (x_2, y_2)$, vertices of G, be joined by the path

$$x = u_0, u_1, \ldots, u_k = y.$$

Due to the definition of a cartesian product, each u_i is obtained from the previous one by a replacement of either coordinate. It is like G_1 and G_2 were busy throwing pies at each other while the target stands still (as in old comic movies). The vertex y represents the state of the fighters after a given number of hits. This means that there are two 'underlying' paths

$$x_1 = x^0, x^1, \ldots, x^{h_1} = x_2$$
$$y_1 = y^0, y^1, \ldots, y^{h_2} = y_2$$

in G_1, G_2, respectively. This implies that $d(x, y) = h_1 + h_2 \geq k_1 + k_2$.

Moreover, if

$$x_1 = x^0, x^1, \ldots, x^{h_1} = x_2$$
$$y_1 = y^0, y^1, \ldots, y^{h_2} = y_2$$

are shortest paths, then

$$(x^0, y^0), (x^1, y^0), \ldots (x^{k_1}, y^0), (x^{k_1}, y^1), \ldots (x^{k_1}, y^{k_2})$$

is a path from x to y. Therefore $d(x, y) \leq k_1 + k_2$ and so $d(x, y) = k_1 + k_2$.

In order to count the number of shortest paths between x and y, let us consider that every such P is obtained by combining a shortest path P_1 between x_1 and x_2 in G_1 and a shortest path P_2 between y_1 and y_2 in G_2, by step-by-step changes of coordinates. Then P is determined by the choice of the steps where P_1 is concerned. Because there are $\binom{k_1 + k_2}{k_1}$ such choices, the result follows. □

Lemma 6.6 *Let G be the cartesian product of the two graphs G_1 and G_2. For $a = (a_1, a_2)$, $b = (b_1, b_2) \in G$, the interval $I(a, b)$ is the (set-theoretic) cartesian product of $I(a_1, b_1)$ and $I(a_2, b_2)$.*

Proof Let us first show that $I(a, b) \subseteq I(a_1, b_1) \times I(a_2, b_2)$. Pick $x = (x_1, x_2) \in I(a, b)$. From the proof of Lemma 6.5, we deduce that any shortest path from a to b is obtained by intertwining two shortest paths of G_1 and G_2, from a_1 to b_1 and from a_2 to b_2, respectively. Then x_1 and x_2 must belong to such paths. This implies $x_1 \in I(a_1, b_1)$ and $x_2 \in I(a_2, b_2)$; therefore $x \in I(a_1, b_1) \times I(a_2, b_2)$. Conversely, assume that $x = (x_1, x_2)$ is such that $x_1 \in I(a_1, b_1)$ and $x_2 \in I(a_2, b_2)$. Due to the aforementioned property of paths, we have $d(a, x) \leq d(a_1, x_1) + d(a_2, x_2)$ and $d(x, b) \leq d(x_1, b_1) + d(x_2, b_2)$. Summing these inequalities, by the hypothesis on x_1 and x_2, we get $d(a, x) + d(x, b) \leq d(a_1, b_1) + d(a_2, b_2) = d(a, b)$. By the triangular inequality it follows that $d(a, x) + d(x, b) = d(a, b)$, that is, $x \in I(a, b)$ as claimed. $\qquad\square$

Theorem 6.7 *Both the cartesian product and the strong product are connectedness preserving.*

Proof For the cartesian product see Exercise 6.4. Moreover, for the strong product, the connectedness of the factors is a necessary condition for the product to be connected. Conversely, since $A(G \square H) \subseteq A(G \boxtimes H))$ and the cartesian product is connectedness preserving, the condition is also sufficient. $\qquad\square$

The lexicographic product $G \propto H$ satisfies $A(G \boxtimes H) \subseteq A(G \propto H))$, with a few cases where equality holds (see Exercise 6.7). However, this operation is not connectedness preserving. Namely, the lexicographic product is *always* connected, regardless of the second factor, provided the first factor is connected.

6.4 More products

A less known but very interesting operation has been introduced by Vizing in 1974. He called it the *modular product*. Actually, following the book of Zykov [267], we will replace it with its complement. However, complements of products are also products in our sense.

Let G and H be graphs. The *modular product* of G and H is $G \lozenge H = G \times H \cup \overline{G} \times \overline{H}$.

Clearly, no clique of $G \lozenge H$ can have order larger than that of either G or H. (A *clique* is a set S of vertices that induces a complete subgraph; when S has k

vertices, we shall call S a k-clique.) The threshold value for these cliques has a special significance, as shown by the following result. A *monomorphism* from a graph G to a graph H is an injective function $f : V(G) \to V(H)$ such that, if a, b are adjacent in G, then $f(a), f(b)$ are also adjacent in H.

Theorem 6.8 *Let G and H be graphs, and let n be the order of G. The number of monomorphisms from G to H is equal to the number of n-cliques of $G \lozenge H$.*

Proof Let $f : V(G) \to V(H)$ be a monomorphism. Consider the set S of all pairs $(v, f(v))$. If uv is an edge of G, then $f(u)f(v)$ must be an edge of H; hence $(u, f(u))$ and $(v, f(v))$ are adjacent in $G \times H$ and so in $G \lozenge H$. On the other hand, if uv is not an edge of G, then $f(u)f(v)$ also is not an edge of H and so $(u, f(u))$ and $(v, f(v))$ are adjacent in $\overline{G} \times \overline{H}$, and hence in $G \lozenge H$.

This proves that the elements of S are pairwise adjacent, so S is an n-clique of $G \lozenge H$.

Conversely, let S be an n-clique of $G \lozenge H$. The first coordinates of its elements run over all of $V(G)$. By associating to each such v the second element v' of the pair $(v, v') \in S$, we get a map $f : v \mapsto v'$ from $V(G)$ into $V(H)$. Reasoning as earlier, we can therefore conclude that f is a monomorphism. $\qquad\square$

This sort of product will not be connectedness preserving in general. On the other hand, it is also predictable that the product of connected non-bipartite graphs is connected, apart from a few exceptions. This is the case for the Knight product, which we shall discuss now.

The classical 8×8 chessboard has always been fascinating even for mathematicians because of the paradoxical clash of the seeming simplicity of the combinatorial problems arising from it and the nonobvious nature of their solutions. Perhaps the most ancient one was that of 'rice-cursive' doubling (from an Indian tale). Albeit elementary as to the mathematics involved (the result is nothing but $2^{128} - 1$), it provides a prime example of unexpected deception of common sense. Combinatorial growth is instinctively underestimated, and this is one of the cultural reasons that delayed the development of discrete mathematics. On a less trivial example, the 'Eight Queens' problem is also classical.

The link of the chessboard with graph theory is fairly natural. The most natural way to describe this object is to consider the cartesian product of two copies of P_7. This is essentially done by the standard international notation, when labelling $a, b, \ldots, h$ the vertices of the first graph and $1, 2, \ldots, 8$ those of the second. However, the rules for moves are not the same for different pieces. They do not always move along paths of that graph: see Exercise 6.8. In view of this exercise, all pieces have 'their' graph, which turns out to be a cartesian or direct product, with the peculiar exception of the Knight.

This one piece can be treated by introducing an ad hoc product, namely, the Knight product defined later. Of course, the graph will be connected because a Knight can walk the whole chessboard.

More generally, it is to be expected that board games and similar maps (such as those created to support many computer games) can often be described as generalised products of graphs.

A remark that will be used later: If G is not bipartite, then G^2 is not bipartite. In fact, a chordless cycle of odd order in G gives rise to a cycle of the same length in G^2. Minimal odd cycles are chordless, so they exist.

The *Knight product* of two graphs or digraphs G, H is defined as follows:

$$G * H = (G \times H^2) \cup (G^2 \times H).$$

Note that if $G = H = P_7$, then the result is precisely the graph whose edges give the possible moves of a Knight on an 8×8 chessboard.

Lemma 6.9 $P_h * P_k$ *is connected if and only if* $h + k \geq 5$.

Proof A straightforward check proves that $P_3 * P_2$ is connected. For all remaining cases with $h + k \geq 5$, the result is easily extended, as any two vertices can be joined by a path of subgraphs isomorphic to $P_3 * P_2$. This proves that the condition is sufficient for the Knight product to be connected. Conversely, assume that $h + k < 5$. Now $P_2 * P_2$ is isomorphic to $K_1 \cup C_8$, while in all remaining cases one of the two factors has less than three vertices, so the graph has no edges. $\qquad\square$

Theorem 6.10 *If both G and H are connected, then $G * H$ is also connected, except for the following cases:*

(i) *either G or H is K_1;*
(ii) *either G or H is K_2 and the other graph is bipartite;*
(iii) *both G and H are star-graphs.*

Proof Clearly, both (i) and (ii) give rise to a disconnected graph. As to (iii), if u, v are the centres of G, H, respectively, then (u, v) is an isolated vertex of $G * H$. Hence, in case (iii) we also have a disconnected product.

Conversely, assume that none of (i)–(iii) hold. We can further assume that both G and H are bipartite; otherwise, $G \times H$ is contained in $G * H$ and so the latter is connected.

Since (iii) does not hold, let G not be a star-graph. Choose any pair (u, v), (u', v') of vertices of $G * H$. Then u and u' belong to a P_3 of G, while v and v' belong to a P_2 of H. By Lemma 6.9, $P_3 * P_2$ is connected, and this is a subgraph of $G * H$. Hence there is a path of $G * H$ joining (u, v) and (u', v'). This proves that $G * H$ is connected.   □

6.5 Stability and two-fold automorphisms

One particular instance of the direct product has given rise to the notion of stability of a graph. Let G be a graph, and consider the direct product $G \times K_2$. This is sometimes called the *canonical double cover* of G, often denoted for short by CDC(G). We generally let the vertex-set of K_2 be $\{0, 1\}$ in this context, therefore the vertices of $G \times K_2$ are of the form $(v, 0)$ or $(v, 1)$ for vertices v in $V(G)$. This graph is bipartite and its colour-classes are the vertex-sets $V(G) \times \{0\}$ and $V(G) \times \{1\}$. It is clear that the automorphism group of $G \times K_2$ contains the direct product of Aut(G) with $\mathbb{Z}_2$. In [175] a graph was defined as unstable if $|\text{Aut}(G \times K_2)| > 2|\text{Aut}(G)|$; G is stable if equality holds.

While in general it is not obvious how to figure out whether a connected graph G is stable or not, there are three cases of 'trivially unstable' graphs, where G is a priori known to be unstable. When G is bipartite the graph $G \times K_2$ consists of two copies of G, therefore bipartite graphs having a nontrivial automorphism are always unstable. Also, if G is disconnected and two components are isomorphic, then G is certainly unstable. Finally, if two vertices u and v of G have the same neighbourhood, then the automorphism of $G \times \mathbb{Z}_2$ that swaps $(u, 0)$ and $(v, 0)$ but fixes $(u, 1)$ and $(v, 1)$ is never a lifting of an automorphism of G. Therefore this is another situation where G is always unstable. Hence, the cases of unstable graphs which we shall consider to be interesting or non-trivial are those which are connected and non-bipartite and do not have a pair of vertices with the same neighbourhood.

Figure 6.2 gives an example of such an unstable graph G together with $G \times K_2$. In order to avoid cluttering the diagram, the vertices $(v, 0)$ and $(v, 1)$ are denoted by v and v', respectively. One can easily check that the automorphism group of G has size 8 while that of $G \times K_2$ has order $2 \times 4! = 48$, therefore G is unstable. We shall, in this section, try to investigate, for an unstable graph G, what sort of 'hidden' symmetries, apart from its automorphisms, G might have which cause the automorphism group of $G \times K_2$ to contain automorphism which are not automorphisms of G lifted by $\mathbb{Z}_2$.

Therefore we shall start by defining what we mean by a two-fold automorphism of a graph. It is important here to recall that we consider every edge

G: $G \times K_2$:

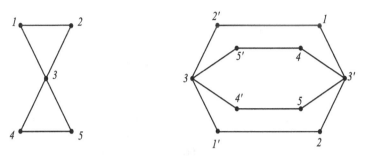

Figure 6.2. A graph G and its canonical double cover

$\{u, v\}$ of a graph to consist of the pair of arcs (u, v) and (v, u). Let α, β be two permutations of $V(G)$ such that (u, v) is an arc of G if and only if $\alpha(u), \alpha(v))$ is an arc of G. In this case, the pair (α, β) is said to be a *two-fold automorphism* of G, which we also call a *TF-automorphism* for short. When α is an automorphism of G then (α, α) is clearly also a TF-automorphism. Vice versa, if $\alpha = \beta$ then α is an automorphism of G. We shall be mostly interested in TF-automorphisms in which $\alpha \neq \beta$; we call such TF-automorphisms *nontrivial*. The set of all TF-automorphisms of G forms a group which we denote by $\text{Aut}^{\text{TF}}(G)$. Therefore, by identifying every automorphism α of G with the TF-automorphism (α, α) we can consider $\text{Aut}(G)$ to be a subgroup of $\text{Aut}^{\text{TF}}(G)$.

For example, if $\alpha = (1\ 4)$ and $\beta = (2\ 5)$ then (α, β) is a nontrivial TF-automorphism of the graph G in Figure 6.2. This can easily be checked because the pair (α, β) exchanges the arcs $(1, 2)$ and $(4, 5)$, takes all arcs incident with 3 into arcs incident with 3, and fixes the arcs $(2, 1)$ and $(5, 4)$.

Consider $\text{Aut}(G \times K_2)$. Let Σ be the set-wise stabiliser of $V(G) \times \{0\}$ in $\text{Aut}(G \times K_2)$, which of course coincides with the set-wise stabiliser of $V(G) \times \{1\}$. We will show that it is the structure of Σ which essentially determines whether $G \times K_2$ has automorphisms which cannot be lifted from automorphisms of G. The following result, which is based on [176], Lemma 2.1, and which can be found in [143] implies that these automorphisms of $G \times K_2$ arise if the action of $\sigma \in \Sigma$ on $V(G) \times \{0\}$ is not mirrored by its action on $V \times \{1\}$. We shall, for short, denote $V(G), V(G) \times \{0\}, V(G) \times \{1\}$ by V, V_0, V_1, respectively.

Lemma 6.11 *Let* $f : \Sigma \to S_V \times S_V$ *be defined by* $f : \sigma \mapsto (\alpha, \beta)$ *where* $(\alpha(v), 0) = \sigma(v, 0)$ *and* $(\beta(v), 1) = \sigma(v, 1)$, *that is* α, β *extract from* σ *its action on* V_0 *and* V_1 *respectively. Then:*

(i) f is a group homomorphism;

(ii) f is injective and therefore $f : \Sigma \to f(\Sigma)$ is a group automorphism;

(iii) the set $f(\Sigma)$ is equal to

$$\{(\alpha, \beta) \in S_V \times S_V : \{x, y\} \in E(G) \Leftrightarrow \{\alpha(x), \beta(y)\} \in E(G)\},$$

that is $f(\Sigma) = \text{Aut}^{TF}(G)$, that is every (α, β) is a TF-*automorphism* of G.

Proof The fact that f is a group homomorphism, that is, that f is a structure-preserving map from Σ to $S_V \times S_V$, follows immediately from the definition since for any σ_1, $\sigma_2 \in \Sigma$ where $f(\sigma_1) = (\alpha_1, \beta_1)$ and $f(\sigma_2) = (\alpha_2, \beta_2)$, $f(\sigma_1)f(\sigma_2) = (\alpha_1\beta_1)(\alpha_2\beta_2) = (\alpha_1\alpha_2, \beta_1\beta_2) = f(\sigma_1\sigma_2)$. This map is clearly injective and therefore $f : \Sigma \to f(\Sigma)$ is a group automorphism.

Consider an arc $((u, 0), (v, 1))$. Then note that since $\sigma \in \Sigma \subseteq \text{Aut}(G \times K_2)$, $(\sigma(u), 0), (\sigma(v), 1))$ is also an arc of $G \times K_2$. By definition, this arc may be denoted by $((\alpha(u), 0), (\beta(v), 1))$ and, following the definition of $G \times K_2$, it exists if and only if $(\alpha(u), \beta(v))$ is an arc of G. Hence f maps elements of Σ to (α, β) which clearly take arcs of G to arcs of G. This implies that (α, β) is a TF-automorphism of G and hence $f(\Sigma) \subseteq \text{Aut}^{TF}(G)$.

Conversely, let $(\alpha, \beta) \in \text{Aut}^{TF}(G)$. Define σ by $\sigma(v, 0) = (\alpha(v), 0)$ and $\sigma(v, 1) = (\beta(v), 1))$, then $f(\sigma) = (\alpha, \beta)$. Hence, $\text{Aut}^{TF}(G) \subseteq f(\Sigma)$. Therefore $f(\Sigma) = \text{Aut}^{TF}(G)$. □

And this is the main theorem of this section, which was first given in [143].

Theorem 6.12 *Let G be a graph. Then*

$$\text{Aut}(G \times K_2) = \text{Aut}^{TF}(G) \rtimes \mathbb{Z}_2.$$

Hence, G is unstable if and only if it has a nontrivial TF-automorphism.

Proof From Lemma 6.11, $f(\Sigma) = \text{Aut}^{TF}(G)$, which must have index 2 in $\text{Aut}(G \times K_2)$. The permutation $\delta(v, \varepsilon) \mapsto (v, \varepsilon + 1)$ is an automorphism of $G \times K_2$ and $\delta \notin f(\Sigma)$. Then $\text{Aut}(G \times K_2)$ is generated by $f(\Sigma)$ and δ. Furthermore, $f(\Sigma) \cap \langle \delta \rangle = \text{id}$ and $f(\Sigma) \lhd \text{Aut}(G \times K_2)$ being of index 2.

Since $\text{Aut}(G \times K_2) = \text{Aut}^{TF}(G) \rtimes \mathbb{Z}_2$, G is stable if and only if $\text{Aut}^{TF}(G) = \text{Aut}(G)$. □

This result can be compared with Theorem 6.1. There, we had that the group $\text{Aut}(G * H)$ contains a subgroup isomorphic to $\text{Aut}(G) \times \text{Aut}(H)$. Here, for the special case of $H = K_2$ and $*$ the direct product, we are given exactly what $\text{Aut}(G * H)$ is.

Going back to Figure 6.2 we deduce that the number of nontrivial TF-automorphisms of G is equal to $|\text{Aut}(G \times K_2)|/2 - |\text{Aut}(G)| = \frac{48}{2} - 8 = 16$. We have given, earlier one nontrivial TF-automorphism of G. Following the proof of Theorem 6.12, the two automorphisms of $G \times K_2$ which are lifts of this nontrivial automorphisms are therefore σ_1 and σ_2 defined as follows:

$$\sigma_1 = (1\ 4)(2'\ 5')$$

and σ_2 is obtained from σ_1 by changing the primed vertices to unprimed, and vice versa, that is,

$$\sigma_2 = (1'\ 4')(2\ 5).$$

As indicated in the proofs given earlier, σ_1 is obtained by letting $\sigma_1(x) = y$ if $\alpha(x) = y$ and $\sigma_1(x') = y'$ if $\beta(x) = y$; analogously, σ_2 is obtained by letting $\sigma_2(x) = y$ if $\beta(x) = y$ and $\sigma_2(x') = y'$ if $\alpha(x) = y$.

6.6 Additional remarks on graph products

We wonder if in some sense this approach to graph products is the 'natural' one, able to cover every special case that reasonably deserves the name of graph (or digraph) product.

In their book, Jensen and Toft [117] give plenty of details about several graph products from the standpoint of graph colourings. There, they notice (p. 181) that previous attempts to give an extensive definition, such as the 2^8 products studied by Nešetřil and Rödl [206], cannot include all known products having $V(G) \times V(G')$ as their carrier. As a counterexample they present the wreath product. Recall that the wreath product of G and G' is defined as the graph on $V(G) \times V(G')$, where (x, y) and (x', y') are adjacent if either $x = y$ and $x'y' \in E(G')$ or $xy \in E(G)$ and $\varphi(x') = y'$ for some automorphism φ of G'.

However, the wreath product could be conveniently defined by our method. Namely, for every set V and self-map φ of V let V_φ denote the digraph on V whose arc-set is $\{(x, y) : \varphi(x) = y\}$. Then the wreath product of G and G' is the union of $G^\circ \times G'$ and all $G \times V(G')_\varphi$ for each $\varphi \in \text{Aut}(G')$. (It turns out to be undirected because the inverse of an automorphism is an automorphism.)

6.7 Exercises

6.1 We have already shown in Section 6.1 that the Petersen graph can be obtained as a product of a graph G of order 5 and a graph H of order 2. Prove that it is not possible to obtain the Petersen graph as a product of G and H in the sense defined in this chapter if, for all TGFs f_i', the graph $f_i'(H)$ is loop-free.

6.2 Give an example of a product $*$ such that an appropriate choice, but not all choices, of disconnected graphs G and H will give a connected $G * H$.

6.3 Define $G * H = (G^2 \times H) \cup (G^3 \times H)$. Show that this product is connectedness preserving but the arc-set $A(G*H)$ does not necessarily contain the arc-set $A(G \times H)$.

6.4 Prove that the cartesian product is connectedness preserving.

6.5 Provide examples of pairs G, G' of nonisomorphic graphs such that $G \times K_2$, $G' \times K_2$ are isomorphic. (See also [230]).

6.6 The definition of the canonical double cover can easily be extended to digraphs (see [213]). Let G be a digraph and let $\sigma \in S_2$ be the transposition (0 1). Define the canonical double cover of G to be the digraph with vertex-set $V(G) \times \{0, 1\}$ and where there is an arc from vertex (u, i) to vertex (v, j) if and only if (u, v) is an arc in G and $\sigma(i) = j$. Show that the canonical double cover of G is isomorphic to the direct product $G \times K_2$.

6.7 Describe the pairs (G, H) such that $G \boxtimes H = G \propto H$.

6.8 The chessboard can be considered as a cartesian product of the two sets $\{a, b, \ldots, h\}$ and $\{1, 2, \ldots, 8\}$. Show that to each piece of the game of chess there could be associated a suitable product of graphs or digraphs defined on these sets, in such a way that every arc xy represents a possible move of the piece from x to y.

6.9 The following construction presents an unstable graph G of any order greater or equal to 6. Let $V(G) = \{1, 2, \ldots, n\}$ where $n \geq 6$. Its edges are: $\{3, 1\}, \{3, 2\}, \{3.4\}$, $\{3, 5\}, \{1, 2\}, \{4, 5\}$ plus all edges of the form $\{1, i\}$ and $\{4, i\}$ for $i = 6, 7, \ldots, n$. Show that the pair of permutations (α, β) where $\alpha = (1\ 4)$ and $\beta = (2\ 5)$ is a TF-automorphism of G, which is therefore a nontrivial unstable graph whose complement is also connected.

6.10 The first case of the previous construction gives the graph G shown in Figure 6.2, which is the smallest non-trivial unstable graph which also has a connected complement. Here G consists of the cycle on the four vertices $1, 3, 4, 6$ in that order, together with a vertex 2 adjacent to vertices 1 and 3 and vertex 5 adjacent to 3 and 4. Clearly, Aut(G) has size 2. Check that Aut($G \times K_2$) has order 8, therefore G is unstable. The automorphism group of $G \times K_2$ is generated by the permutations $\sigma_1 = (1\ 4)(2'\ 5'), \sigma_2 = (1'\ 4')(2\ 5)$ and the involution interchanging each i with i'. Note that neither σ_1 nor σ_2 can be lifts of automorphisms of G.

6.11 Another small, connected, non-bipartite, unstable graph G having a connected complement is shown in Figure 6.4.

    (a) Show that $|\text{Aut}(G)| = 4$. Consider graph H also shown in Figure 6.4. Show that $|\text{Aut}(H)| = 8$ and that $H \times K_2 \simeq G \times K_2$. Deduce that G is unstable.

    (b) Note that $\sigma = (2\ 6)(3'\ 5')$ is an automorphism of $G \times K_2$ which cannot come from a lifting of an automorphism of G because its action on one colour class of $G \times K_2$ is not mirrored by its action on the other colour class. In fact, it is the lift of the TF-automorphism (α, β) on G where $\alpha = (2\ 6)$ and $\beta = (3\ 5)$.

    (c) Determine $|\text{Aut}(H \times K_2)| = |\text{Aut}(G \times K_2)|$ and deduce that H is stable.

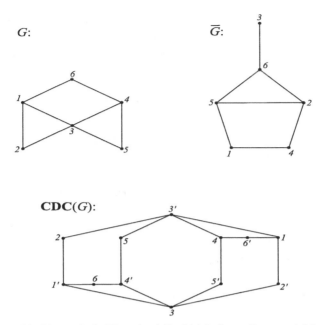

Figure 6.3. The graph G of Exercise 6.10 which is the smallest nontrivial unstable graph which also has a connected complement

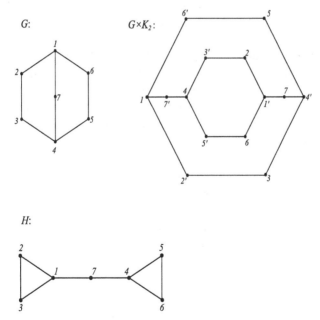

Figure 6.4. An illustration of Exercise 6.11: two graphs G and H having the same CDC

6.12 The canonical double cover of the Petersen graph is the well-known Desargues Graph. Show that the Petersen graph is stable. Find a graph G which has the same canonical double cover as the Petersen graph. Show that G is not stable.

6.8 Notes and guide to references

Parts of this chapter are presumably known as 'folklore', in particular where the exercises are concerned. Definitions are well known and easy to find in many textbooks; this is why our list of references pertaining to this chapter is fairly short.

Actually, graph products arise in several papers, albeit they are mostly considered as a tool to build examples rather than for their own sake.

Focus on properties of such operations has been considered, though. Other than the pioneering work by Sabidussi [224], which focused on the infinite case, several further investigations are recalled in [95] as well as in the papers mentioned there. We have not gone through the details here because the article itself is of a survey nature.

In both of these papers, as well as for the modular product by Vizing (described in the book [267] by Zykov), attention is paid to specific rather than general graph operations. It seems that other authors are gaining interest in finding ways of defining new products; see, for example, [234] and [1].

A different approach, closer to the general viewpoint adopted here, has been taken by Imrich and Izbichi in [113]. There, associative products of undirected graphs (with no twisting) are characterised. See also [110] and [207].

We have very briefly treated the concept of stability of a graph G which arises from consideration of the direct product of G with K_2. Many more results on stability of graphs can be found in [215, 213, 175, 176, 260, 240, 239, 241, 143] for example, including results investigating when two nonisomorphic graphs can have the same canonical double cover, an example of which is discussed in Exercise 6.11. The graph G of Exercise 6.12 which has the same canonical double cover as the Petersen graph was, as far as we know, first published in [215]. In this chapter we have concentrated on the relatively new idea of explaining instability in terms of two-fold automorphisms. Much more detail about two-fold automorphisms can be found in [141, 142, 143, 145, 144].

A treatment of k-coverings which generalises that given in Exercise 6.6 can be found in [24].

The book [114] gives an exhaustive coverage of the four main graph products: cartesian, direct, strong and lexicographic. The more recent handbook [96] gives an exhaustive coverage of graph products and the many ramifications of their theory and applications.

7

Special Classes of Vertex-Transitive Graphs and Digraphs

The class of Cayley digraphs far from exhausts all vertex-transitive digraphs. The same holds even for the undirected case: The smallest example of a non-Cayley vertex-transitive graph is the Petersen graph; we have already seen that this is not a Cayley graph.

In an earlier chapter we discussed the problem of obtaining non-Cayley vertex-transitive graphs, and we have shown one simple, systematic way of obtaining them. Such constructions often lead to very interesting classes of graphs. The size of the automorphism group is generally not a good indicator for deciding whether a vertex-transitive graph is Cayley, in the sense that a graph may be rich with automorphisms and be non-Cayley, or have a small automorphism group and be Cayley. Examples for the former situation are provided by the odd graphs. The latter case arises with GRRs: in this case, we have a paradox because the smallest possible size for $\mathrm{Aut}(G)$ guarantees that the graph is Cayley.

The five classes of graphs that we are going to introduce in this chapter (the generalised Petersen graphs, Kneser graphs, metacirculant graphs, quasi-Cayley digraphs and generalised Cayley graphs) are essentially different. Each of these classes contains infinitely many further non-Cayley graphs and digraphs and some are not even vertex-transitive.

While the methods of orbital graphs and double-cosets representations, as described in the previous chapters, enable us to represent all possible vertex-transitive graphs, a general classification theorem for them is presently far beyond us.

Special cases of vertex-transitive graphs have been classified. It is known that all those of prime order are Cayley (actually, *circulant* graphs, that is, Cayley graphs of a cyclic group). A classification result has also been obtained for the next case, that of order equal to the product of two primes. This required the help of the classification theorem for finite simple groups and further deep group-theoretical results. Further details are given in Section 7.3.

In view of these remarks, this chapter is not intended to offer classification theorems, but rather to illustrate several significant classes of graphs having a good symmetry (mostly, vertex-transitivity).

Each of the next sections is devoted to one of these classes, collecting results that are mostly available only in journal articles.

7.1 Generalised Petersen graphs

Let $n > k \geq 1$. The graph $G = \text{GPG}(n,k)$ having $V(G) = \{u_i \mid i \in \mathbb{Z}_n\} \cup \{v_i \mid i \in \mathbb{Z}_n\}$ as vertex-set and with exactly the following adjacencies for all i:

(i) u_i is adjacent to v_i;

(ii) u_i is adjacent to v_{i+1};

(iii) v_i is adjacent to v_{i+k};

is called a *generalised Petersen graph*.

Clearly $\text{GPG}(n,k) \cong \text{GPG}(n, n-k)$ and $\text{GPG}(n,1) \cong C_n \square K_2$. Note that $\text{GPG}(5,2)$ is the Petersen graph.

In general, $G = \text{GPG}(n,k)$ has the automorphisms f and g defined by:

$$f : u_i \mapsto u_{i+1}, \quad v_i \mapsto v_{i+1},$$

$$g : u_i \mapsto u_{-i}, \quad v_i \mapsto v_{-i}.$$

Henceforth, $\text{Aut}(G)$ will be assumed to contain the dihedral subgroup $\langle f, g \rangle \simeq D_n$. This is, of course, a metacyclic group, but never transitive: its orbits are $\{u_i \mid i \in \mathbb{Z}_n\}$ and $\{v_i \mid i \in \mathbb{Z}_n\}$. Unlike the Petersen graph, many $\text{GPG}(n,k)$ have no further automorphisms other than the elements of $\langle f, g \rangle$, so that they are not vertex-transitive. Namely, if $(n,k) \neq 1$, the set of v_i splits into (n,k) cycles, each of length $\frac{n}{(n,k)}$ (here, (n,k) denotes the highest common factor of n and k), while the u_i form a unique cycle.

The following result of Frucht, Graver and Watkins [80] characterises the vertex-transitive generalised Petersen graphs. They also showed in this paper that further properties, like edge-transitivity, hold only for a finite number of pairs (n,k).

Theorem 7.1 (Frucht, Graver and Watkins) *The graph* $\text{GPG}(n,k)$ *is vertex-transitive if and only if* $k^2 \equiv \pm 1 \pmod{n}$, *or else* $(n,k) = (10,2)$.

The exceptional case GPG(10, 2) is the dodecahedron, which is also arc-transitive but non-Cayley. In fact, the following result proved by Nedela and Škoviera [204] and by Lovrečič Saražin [154] holds.

Theorem 7.2 (Nedela and Škoviera; Lovrečič Saražin) *The graph* GPG(n, k) *is Cayley if and only if* $k^2 \equiv 1 (\mathrm{mod}\, n)$.

As was just seen, the usual description of the generalised Petersen graphs is explicit: there are these vertices and these edges, etc. An alternative one is provided here. Namely, Corollary 7.4 characterises them amongst all cubic graphs.

Let us recall that, given a graph $G = (V, E)$, a set $X \subseteq V$ is called *dominating* if, for each vertex $u \in V - X$, there exists $v \in X$ such that u is adjacent to v.

Theorem 7.3 *Let G be a cubic graph, having a dominating cycle $C = (u_0, \ldots, u_{p-1})$, $3 \le p < |V(G)|$ and an automorphism f taking each u_i into u_{i+1} (where addition is modulo p). Then there are vertices v_j such that $V(G) = \{u_i | i \in \mathbb{Z}_p\} \cup \{v_j | j \in \mathbb{Z}_q\}$ and the edge-set of G is one of the following, with no edges other than the ones listed:*

(i) $u_i u_{i+1}$, $u_i v_i$, $v_i v_{i+k}$, *with $p = q$ and $k \le q$ fixed;*

(ii) $u_i u_{i+1}$, $u_j v_j$, $v_j u_{j+n}$, $v_j v_{j+k}$, *with $p = 2q$, $q = 2k$;*

(iii) $u_i u_{i+1}$, $u_i v_{[i]}$, *and $p = 3q$, where $[i]$ is i mod q.*

Conversely, each graph described by one of (i)–(iii) *satisfies this condition with $C = (u_0, \ldots, u_{p-1})$.*

Proof Let v_0 be a vertex not in C, such that v_0 is adjacent to u_0. Under the action of f, we obtain vertices v_i such that $u_i v_i$ are edges, $i \in \mathbb{Z}_m$.

Because C is dominating, we have that $p \ge |V(G)|/2$. If $p = |V(G)|/2$, then the v_i are all distinct. Now v_0 is adjacent to v_k for exactly one k, so that $v_i v_{i+k} \in E(G)$ for all i. Thus (i) holds, with $p = q$.

Assume now that $p > |V(G)|/2$. Let q be the smallest integer such that $v_j = v_{j+q}$ for some j. Clearly, this equality holds for *all* j. If some v_j, and so every v_j, is adjacent only to vertices of C, these are u_j, u_{j+q} and u_{j+2q}. Therefore $p = 3q$, and (iii) follows.

Otherwise, a vertex not in C is adjacent to some other vertex not in C. The subgraph induced by $V(G) - C$ is then a disjoint union of copies of K_2, say qK_2. This easily leads to (ii).

The converse, namely that (i)–(iii) imply the existence of a dominating cycle C with the stated properties, is easy to prove. ◻

Note that for $k = 1$ we get from (ii) the Kuratowski Graph $K_{3,3}$. The smallest instances of graphs corresponding to (iii) are K_4 and the cube Q_3, obtained for $q = 1$ and $q = 2$, respectively.

Corollary 7.4 *A cubic graph with $2p$ vertices ($p \geq 3$) is a generalised Petersen graph if and only if it has a dominating cycle $C = (u_0, \ldots, u_{p-1})$ and an automorphism taking each u_i into u_{i+1}.*

Proof The resulting graph is described by Proposition 7.3(i) and is exactly a generalised Petersen graph.   □

We conclude this section with some informal information about the *Frucht notation*.

Apart from very small graphs, to draw a graph is by no means trivial. To make a picture representing a large one may be time-consuming and almost useless for checking the actual properties of the graph. The idea of Frucht's notation, which will be illustrated by Examples 1 and 2, is to choose an appropriate partition $\{V_1, \ldots, V_s\}$ of the vertex-set so that each V_i induces some 'elementary' graph (e.g. a cycle), and links between different V_i can also be described in a most easy way (e.g. with a labelling as in the generalised Petersen graphs). The output will be a diagram with circles (representing the V_i) linked by edges and arcs. Inside the former and above the latter, symbols are written to represent which kinds of graphs are inside the circles and how they are related.

Examples

1) Let us consider the graph G drawn in Figure 7.1, which is represented by the Frucht notation shown in the same diagram. There are six vertices u_i ($i = 0, \ldots, 5$) on the outer cycle and another six vertices v_j ($j = 0, \ldots, 5$) on the inner cycle of G. These will be represented by the left and the right circles of the diagram, respectively. Moreover, vertices u_i, u_j are adjacent if and only if $i - j = \pm 1$, where addition is taken modulo 6. This situation is represented by the symbol 6(1) inside the left circle of the diagram. Similarly, v_i is adjacent to v_j if and only if $i - j = \pm 1$ or $i - j = \pm 2$. This corresponds to the symbol 6(1, 2) inside the right circle. The edge between the two circles means that u_i is adjacent to v_j whenever $i \equiv j \pmod 6$. The arrow from left to right, labelled by 1, indicates that u_i is adjacent to v_j whenever $j \equiv i + 1 \pmod 6$.

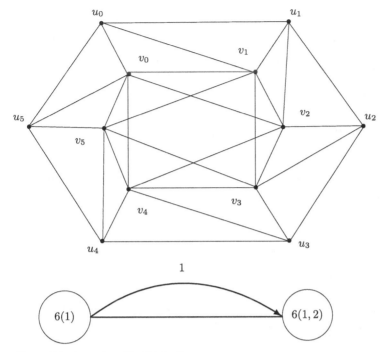

Figure 7.1. A graph and its Frucht diagram

Figure 7.2. The Frucht diagram of the generalised Petersen graph

This kind of description is most effective when a semiregular automorphism f can be found (a permutation is called *semiregular* if its decomposition into disjoint cycles gives cycles all of the same length). Then we take $\{V_1, \ldots, V_s\}$ as the set of all orbits of f. Now each V_i induces a circulant graph (for the restriction of f to this graph is transitive), and f can lead the description of edges between different V_i.

2) The graph GPG(n, k) has the semiregular automorphism

$$f = (u_0 \cdots u_{n-1})(v_0 \cdots v_{n-1}).$$

Hence Figure 7.1 represents GPG(n, k).

7.2 Kneser graphs and odd graphs

Let k and n be positive integers with $n \geq 2k+1, k \geq 2$. We define $K = K(n,k)$ as the graph with the k-subsets of $\{1,2,\ldots,k\}$ as vertices, where two such vertices are adjacent if they are disjoint as sets. These graphs are commonly referred to as *Kneser's graphs*. If $n = 2k+1$, then $K(n,k)$ is also referred to as the *odd graph* O_{k+1}.

The following theorem of Godsil provides the conditions on k which imply that the corresponding Kneser graphs are Cayley graphs.

Theorem 7.5 (Godsil) *Except in the following cases, $K(n,k)$ is not a Cayley graph.*

(i) *$k = 2$, n is a prime power and $n \equiv 3 \mod 4$;*
(ii) *$k = 3$, $n = 8$ or 32.*

Furthermore, if $n = 2k + 1$ and $k \neq 2$ or 4, a transitive subgroup of $\Gamma = \mathrm{Aut}(K)$ is isomorphic (as an abstract group) to S_n or A_n.

Proof It is not difficult to show (Exercise 7.5) that if Γ is a transitive subgroup of $\mathrm{Aut}(K)$ and $\phi \in \Gamma$ and $x, y \in V(K)$, then

$$|x \cap y| = |\phi(x) \cap \phi(y)|.$$

Using this it is easy to show that Γ is just S_n acting on $V(K)$ in the representation induced by its action on $\{1,2,\ldots,n\}$.

As has been proved earlier in this book, a graph G is a Cayley graph for some group if and only if $\mathrm{Aut}(G)$ contains a subgroup acting regularly on $V(G)$. We will show that, except in cases (i) and (ii), Γ contains no regular subgroup.

Let $\mathcal{H}$ be a transitive subgroup of Γ. The isomorphism from Γ to S_n takes $\mathcal{H}$ onto a subgroup $\bar{\mathcal{H}}$ of S_n that acts transitively on the k-subsets of $\{1,2,\ldots,n\}$. Now, if $\bar{\mathcal{H}}$ is actually k-transitive, then $n!/(n-k)!$ divides $|\mathcal{H}| = |\bar{\mathcal{H}}|$. Since $|V(K)| = \binom{n}{k}$, it follows that the vertex stabiliser of $\mathcal{H}$ has order at least $k!$. Therefore, if $\bar{\mathcal{H}}$ is k-transitive, then $\mathcal{H}$ is not regular.

Recall that a group that acts transitively on the k-element subsets of a set is said to be k-homogeneous. By Theorem 2 of [149], when $k \geq 5$, a k-homogeneous group is k-transitive. Consequently, if $k \geq 5$, Γ contains no regular subgroup. Suppose $k < 5$. In this case the k-homogeneous groups that are not k-transitive have been determined by Kantor in [119]. Comparing the orders of these groups with $\binom{n}{k}$ yields the conclusion that, unless (i) or (ii) holds, Γ contains no regular subgroup.

Assume now that $n = 2k + 1$. Let $\mathcal{H}$ be a transitive subgroup of Γ and let $\bar{\mathcal{H}}$ be defined as earlier. Then $\bar{\mathcal{H}}$ is k-homogeneous and so, by Theorem 2 of

[149], it is j-homogeneous for all j such that $1 \le j \le k$. Furthermore, if $\bar{\mathcal{H}}$ is j-homogeneous, then it is $(n - j)$-homogeneous. Thus $\bar{\mathcal{H}}$ is j-homogeneous for all j, $1 \le j \le n-1$. By the results of Beaumont and Peterson [21], we conclude that either $\mathcal{H}$ contains the alternating group A_n (and so $|\Gamma : \mathcal{H}| \le 2$) or $k = 2$ or 4. $\qquad\square$

Note that when $k = 2$, $K(n, k)$ is isomorphic to the complement of the line graph of K_n.

7.3 Metacirculant graphs

Let $m, n \ge 1$ be integers. A graph G is said to be (m, n)-*metacirculant* graph if it has an automorphism f with a cycle decomposition

$$ f = \left(v_0^0 v_1^0 \cdots v_{n-1}^0 \right) \left(v_0^1 v_1^1 \cdots v_{n-1}^1 \right) \cdots \left(v_0^{m-1} v_1^{m-1} \cdots v_{n-1}^{m-1} \right) $$

and an automorphism g that cyclically permutes the orbits

$$ V_i = \left\{ v_0^i, v_1^i, \ldots, v_{n-1}^i \right\} \quad (i = 0, 1, \ldots, m - 1). $$

It can be shown that there exists $r \in \mathbb{Z}_m^*$ (the group of all elements of $\mathbb{Z}_m$ having a multiplicative inverse mod m) such that $g(v_j^i) = v_{rj}^{i+1}$ for all i and j, forcing $r^h \equiv 1 \mod n$ for some multiple h of m. This means that a metacirculant graph is a graph with a transitive cyclic or metacyclic automorphism subgroup. Note that this terminology is consistent because in the first case the graph is circulant. It is proved in [8] that every (m, n)-metacirculant graph can be associated with an array $(m, n, r, S_0, S_1, \ldots, S_k)$, where $k = \lfloor \frac{m}{2} \rfloor$, $S_i = \left\{ s \in Z_n : v_0^0 v_s^i \text{ is an edge} \right\}$ and $r^m S_i = S_i$. Let

$$ M(m, n, r, S_0, S_1, \ldots, S_k) $$

denote the corresponding graph.

Metacirculant graphs can also be described by subsets of certain groups, in a way quite similar to Cayley graphs. In fact, circulant graphs are nothing but Cayley graphs of cyclic groups. However, many metacirculant graphs are non-Cayley (the Petersen graph amongst others).

Though these graphs were studied only in the undirected case, their straightforward generalisation to digraphs is possible, as remarked in [174]. When they were described in [174], in order to facilitate the classification of vertex-transitive graphs of order pq, it was found advisable to adopt the following (nonstandard) terminology.

Letting $p \geq q$, a graph G as earlier will be called:

- *p-imprimitive* if there is a transitive $\Gamma \leq \mathrm{Aut}(G)$ having a subgroup with an imprimitivity block system with blocks of order p;
- *q-imprimitive* if it is not p-imprimitive but there is a transitive $\Gamma \leq \mathrm{Aut}(G)$ having a subgroup with an imprimitivity block system with blocks of order q;
- *primitive* if it is neither p-imprimitive nor q-imprimitive, that is, $\mathrm{Aut}(G)$ has definitively no imprimitive subgroup.

Hence the word 'primitive' is used in a more restrictive sense here than usual. For example, the Petersen graph is 5-imprimitive, even though its full automorphism group is primitive. But our choice is reasonable, in view of the fact that the existence of an imprimitive subgroup Γ can be exploited in order to describe the graph, no matter if Γ is contained in a larger, primitive subgroup.

The full proof of Theorem 7.6 is very long. The first step was [165], which took care of the p-imprimitive groups: they are all metacirculant graphs. We will then describe here the remaining cases. Note that the definition of a metacirculant graph implies that every Cayley pq-graph is metacirculant, so all graphs listed in (i)–(iv) are non-Cayley vertex-transitive graphs.

Note that the the following classification involves a new class of graphs introduced in [169] and called here MS-graphs, in the sense of [230].

Theorem 7.6 *A vertex-transitive pq-graph must be one of the following graphs:*

- (i) *a metacirculant graph;*
- (ii) *an MS-graph with p a Fermat prime and q | p − 1;*
- (iii) *an orbital graph arising from certain rank 3 representations of $P\Omega^{\pm}(2d, 2)$ or M_{22};*
- (iv) *a generalised orbital graph associated with the action of A_7 on triples;*
- (v) *a generalised orbital graph associated with a two-dimensional projective special linear group.*

Furthermore, these five classes of graphs are mutually disjoint.

Further attempts have been carried out to extend this result to the product of more than three primes (see, for example, [177]).

To illustrate some of the typical reasonings performed when dealing with this sort of result, let us show how the classification is carried out in the much simpler case of prime order.

Theorem 7.7 *All vertex-transitive graphs and digraphs of prime order are Cayley graphs on cyclic groups (that is, circulant graphs).*

Proof Assume that G is a vertex-transitive graph or digraph of prime order p. Its automorphism group Γ must have order a multiple of its degree, that is p. Therefore, Sylow's Theorem guarantees the existence of an element π of order p in Γ.

Now, π must be a product of cycles of prime order p (plus possibly fixed points). However, the degree of Γ is already p, thus π necessarily consists of a single cycle of order p only. Then the cyclic group generated by π has order p and is transitive, and hence regular. $\qquad\square$

7.4 The quasi-Cayley graphs and digraphs

In order to generalise the notion of a Cayley digraph, we may note that we do not need a group in the definition, but any groupoid (i.e. a set with any binary operation) would work, except that not very much can be expected for the symmetry of such a graph. But special kinds of groupoids are still of interest. Recall that a *quasigroup* is a groupoid $(Q, \cdot)$, where for each choice of $a, b \in Q$ the equations $ax = b$ and $xa = b$ both have a unique solution in Q. In other words, the maps $x \mapsto ax$ and $x \mapsto xa$ are permutations, just as in the case of groups. If, furthermore, $E \subseteq Q$ satisfies $a(bE) = (ab)E$ for all $a, b \in E$, then the quasigroup is said to be *right-associative*.

Let Q be a quasigroup and let $E \subseteq Q$. The *quasigroup digraph* $G = QG(Q, E)$ is defined by $V(G) = Q$, $A(G) = \{(u, ue) \mid u \in Q, e \in E\}$. In particular, if E is right associative, G is said to be a *quasi-Cayley digraph*.

The following result was shown by Dörfler.

Theorem 7.8 (Dörfler) *A digraph G is regular if and only if it is a quasigroup graph. Moreover, if $G = QG(Q, E)$ with E right-associative, then G is vertex-transitive.*

A proof of the second part of Theorem 7.8 can be given by an explicit representation of a quasi-Cayley digraph as a union of orbitals; see Exercise 7.13.

Further properties of quasi-Cayley graphs were later found and exploited in Gauyacq (see [84]), where several examples are provided of vertex-transitive graphs that are not quasi-Cayley. Infinitely many belong to the class of Kneser graphs. In fact, in view of Theorem 7.9, these are seldom Cayley graphs, while in the following theorem she proved that they are all quasi-Cayley.

Following [83], a set F of permutations on a set V is said to be a *regular family* if, for each pair (u, v) of elements of V, there is exactly one $f \in F$ such that $f(u) = v$. Note the similarity of the following theorem and its proof with

Theorem 3.5, although here more care is required because we are dealing with a quasigroup instead of a group.

Theorem 7.9 *Let G be a connected graph. The automorphism group of G contains a regular family on $V = V(G)$ if and only if G is a quasi-Cayley graph.*

Proof ($\Leftarrow$) Let G be $QC(Q, E)$. For any $a \in Q$, the permutation ϕ_0 of Q that maps q into aq is an automorphism of G. Put $F = \{\phi_a : a \in Q\}$. F is a regular family of automorphisms of G on V because, for every pair (u, v) of elements of Q, there exists a unique element $a \in Q$ such that $au = v$.

($\Rightarrow$) Let G be a connected graph and F be a family of automorphisms of G that is regular on V. Let $V = \{1, 2, \ldots, n\}$ and let f_i, $i = 1, 2, \ldots, n$ be the automorphism elements of G such that $f_i(1) = i$. We define in $\{1, 2, \ldots, n\}$ the binary operation $*$ as $i * j = f_i(j)$. The set $\{1, 2, \ldots, n\}$ endowed by with $*$ is a quasigroup Q.

Indeed, the equation $i * x = j$ has a unique solution because f is a permutation of V. The equation $x * i = j$ has a unique solution because there exists a unique f_x in F such that $f_x(i) = j$. Because of the definition of f_i, 1 is the neutral element from the right.

Let E be the set of neighbours of 1 in G. If G is without loops, then $1 \notin E$. Moreover, E is right-associative. Indeed, $(i * j) * E = (f_i(j)) * E$ is the image of E for the automorphism ϕ, which belongs to F, defined by $\phi(1) = f_i(j)$; therefore this is the neighbourhood of $f_i(j)$. Furthermore, $i * (j * E) = f_i(f_j(E))$ is the image by f_i of the neighbourhood of j. Therefore it is the neighbourhood of the $f_i(j)$. If $a \in E$, then the solution of $a * x = 1$ is $f^{-1}(1)$. Since $\{1, a\}$ is an edge of G, the same holds for $\{f_a^{-1}(1), f_a^{-1}(a)\}$ or $f_a^{-1}(a) = 1$; hence $f_a^{-1}(a)$ is an element of E.

Let us now show that E is the generating set of Q. Let

$$E = \{a_1, a_2, \ldots, a_k\}$$

and let $v \in V$. We are going to show, by induction on $d(1, v)$ (the distance between vertices 1 and v), that every $v \in V$ can be expressed as a product of elements of E. If $d(1, v) = 1$, then $v \in E$ by definition of E. Let us assume that all v such that $d(1, v) < k$ can be expressed as a product of a_i. Let u be such that $d(1, u) = k$. There exists a vertex v adjacent to u such that $d(1, v) = k - 1$.

Then,

$$\{u, v\} \text{ is an edge of } G$$
$$\Leftrightarrow \quad \{f_v^{-1}(u), f_v^{-1}(v)\} \text{ is an edge of } G$$
$$\Leftrightarrow \quad \{f_v^{-1}(u), 1\} \text{ is an edge of } G$$
$$\Leftrightarrow \quad \text{there exists } j \in \{1, \ldots, k\} \text{ such that } f_v^{-1}(u) = a_j$$
$$\Leftrightarrow \quad \text{there exists } j \in \{1, \ldots, k\} \text{ such that } u = f_v(a_j) = v * a_j.$$

By the induction hypothesis, v is expressed as a product of a_i; therefore the same holds for u. This is enough to conclude that E is a generating set of Q.

Finally, E has the properties needed to define $QC(Q, E)$, namely,

$$\{u, v\} \text{ is an edge of } G \quad \Leftrightarrow \quad \{1, f_v^{-1}(v)\} \text{ is an edge of } G$$
$$\Leftrightarrow \quad f_u^{-1}(v) \in E \Leftrightarrow v \in f_u(E).$$

Consequently, G and $QC(Q, E)$ are isomorphic. □

7.5 Generalised Cayley graphs

We have already seen (Chapter 6, Section 5) that the direct product of a graph G with K_2, from now on called canonical double cover or CDC of G, and denoted by the symbol CDC(G), may have 'unexpected' automorphisms which do not arise from the lifting of any automorphisms of the starting graph G. It is then natural to ask when the symmetry of CDC(G) has properties that do not hold in G. In particular, in view of the aforementioned discussion, we may expect that CDC(G) can be a Cayley graph even when G is not. The construction discussed in this section, based on (Marušič et al. [176]), gives rise to a large class of graphs whose CDC is a Cayley graph.

In the sequel, the involution of the vertices of CDC(G) that exchanges $(v, 0)$ and $(v, 1)$ for all v in G will be called the *standard automorphism* of CDC(G) and denoted by τ.

Let Γ be a group, let f be an automorphism of Γ, and let A be a subset of Γ. Assume that the following conditions hold:

(i) $f^2 = \mathrm{id}$;
(ii) if $x \in \Gamma$ then $f(x^{-1})x \notin A$;
(iii) if $x, y \in \Gamma$ then $f(x^{-1})y \in A$ implies $f(y^{-1})x \in A$.

The *generalised Cayley graph* $G = \mathrm{GCay}(\Gamma, f, A)$ associated with the triple (Γ, f, A) has Γ as vertex-set, with x adjacent to y whenever $f(x^{-1})y \in A$. Note that (ii) excludes loops in G and (iii) guarantees that G is undirected.

Besides, (iii) implies $f(A^{-1}) = A$, by letting $x = 1$. In view of (i), the condition $f(A^{-1}) = A$ can also replace (iii). Namely, assuming this equality, $f(x^{-1})y \in A$ implies $x^{-1}f(y) \in f(A) = A^{-1}$, and so $f(y^{-1})x = (x^{-1}f(y))^{-1} \in A$.

Clearly, taking f to be the identity automorphism of Γ we get Cayley graphs. By weakening conditions (i)–(iii), we would obtain other kinds of graphs and digraphs, but we shall only consider this definition of generalised Cayley graphs in this section, where all graphs will be simple and undirected.

Let $G = \mathrm{GCay}(\Gamma, f, A)$. Letting z be a fixed element of Γ, consider the permutations $\alpha : x \mapsto zx$ and $\beta : x \mapsto f(z)x$. The pair (α, β) turns out to be a TF-automorphism of G. Namely, (α, β) takes the arc (x, y) into $(zx, f(z)y)$. The condition $f(x^{-1})y \in A$ is tantamount to $f((zx)^{-1}f(z)y = f(x^{-1})f(z)^{-1}f(z)y = f(x^{-1}))y \in A$. So (x, y) is an arc of G if and only if $(zx, f(z)y)$ is.

As proved in [176], if G is a generalised Cayley graph, then $\mathrm{CDC}(G)$ is a Cayley graph. The following class of examples, again taken from [176], shows that a generalised Cayley graph might not be transitive. Let $X(n)$, $n \geq 3$ denote the generalised Cayey graph $GC(\mathbb{Z}_n \times \mathbb{Z}_n, A, \alpha)$ where $A = \{(1, 0), (0, n-1), (1, 1), (n-1, n-1)\}$ and $\alpha : (j, k) \mapsto (k, j)$, for $j, k \in \mathbb{Z}_n$. It is shown in [176] that $X(n)$ is not vertex-transitive.

The following result was already established in [176], but in the light of the concept of TF-automorphisms we can give a much simpler proof.

Proposition 7.1 *Let G be a generalised Cayley graph. If G is stable, then it is a Cayley graph.*

Proof We have observed that each element of the group Γ gives rise to TF-automorphisms (α, β) of G; clearly α, β are different whenever $f \neq \mathrm{id}$. Therefore, if G is stable, then $f = \mathrm{id}$. $\square$

In particular, all non-vertex-transitive generalised Cayley graphs, like the $X(n)$ introduced earlier, are necessarily unstable.

7.6 Exercises

7.1 Let k be relatively prime to n and let t be such that $kt \equiv 1 \bmod n$. Show that $\mathrm{GPG}(n, k) \simeq \mathrm{GPG}(n, t)$.

7.2 Show that the Petersen graph is metacirculant.

7.3 Show that

  (a) $GPG(n, 2)$ is Hamiltonian if and only if $n \not\equiv 5 \bmod 6$;
  (b) $GPG(n, 3)$ is Hamiltonian for $n \neq 5$.

7.4 Determine when $GPG(2k, k)$ is Hamiltonian.

7.5 Show that if ϕ is an automorphism of the Kneser graph $K(n, k)$ and $x, y \in V(K(n, k))$, then $|x \cap y| = |\phi(x) \cap \phi(y)|$.

7.6 Kneser conjectured that: 'If the $\binom{2k+t}{k}$ k-subsets ($n \geq 1, k \geq 1$) of a $(2k + t)$-set are partitioned into $t + 1$ classes, then one of these classes contains two disjoint k-subsets.' Show that this conjecture is equivalent to the statement that: 'The chromatic number of $K(2k + t, k)$ equals $t + 2$.'

7.7 Represent the odd graph O_4 using the Frucht notation.

7.8 Suppose that the graph G has Frucht notation containing a number of 'circles' each containing r vertices, and suppose that these circles are joined by a succession of arrows or lines that begins and ends at the same circle and passes through each circle once (that is, forms a 'cycle' amongst the 'circles'). Suppose also that the sum of the labels on the arrows or lines is relatively prime to r. Show that G is Hamiltonian.

7.9 Show that $\text{Aut}(O_{k+1}) \simeq S_{2k+1}$. [*Hint*: Use the Theorem of Erdös, Ko and Rado, which implies that the independence number of O_{k+1} is $\binom{2k}{k-1}$ and that an independent set with this number of vertices is equal to $\{v \in V(O_{k+1}) : s \in v\}$ for some fixed $s \in \{1, 2, \ldots, 2k + 1\}$.]

7.10 Show that if G is an (m, n)-metacirculant graph as described earlier, then there exists an $r \in \mathbb{Z}_m^*$ such that $g(v_j^i) = v_{rj}^{i+1}$ for all i and j, and that therefore $r^h \equiv 1$ mod n for some h multiple of m.

7.11 Show that every Cayley graph on pq vertices, p, q prime, is metacirculant.

7.12 A famous result due to Burnside says that if (Γ, Y) is a transitive permutation group with $|Y| = p$, a prime, then either Γ is 2-transitive or Y can be identified with the field $\mathbb{Z}_p$ such that Γ is a subgroup of the affine group of transformations of $\mathbb{Z}_p$,

$$\{\sigma_{a,b} : b \in \mathbb{Z}_p^*, b \in \mathbb{Z}_p\},$$

where $\sigma_{a,b}$ is the permutation $x \mapsto ax + b$. Using this result, prove the following.

(a) Let $G = \text{Cay}(\mathbb{Z}_p, S)$. Show that, unless $S = \emptyset$ or $S = \mathbb{Z}_p$,

$$\text{Aut}(G) = \{\sigma_{a,b} : a \in \mathcal{H}, b \in \mathbb{Z}_p^*\},$$

where $\mathcal{H} = \mathcal{H}(S)$ is the largest even order subgroup of $\mathbb{Z}_p^*$ such that S is a union of cosets of $\mathcal{H}$.

(b) Let $G_1 = \text{Cay}(\mathbb{Z}_p, S)$ and $G_2 = \text{Cay}(\mathbb{Z}_p, T)$. Show that $G_1 \simeq G_2$ if and only if there is a positive integer $q \leq (p - 1)/2$ such that $T = qS$. (Compare with Exercise 5.8.)

7.13 Prove the second part of Theorem 7.8 by presenting an explicit representation of G as a quasi-Cayley digraph [*Hint*: Letting M be the permutation group generated by all $x \mapsto gx$ for $g \in Q$, the union of all orbital graphs defined by the pairs $(1, e)$, $e \in E$, is exactly $QG(Q, E)$: It suffices to remark that $(gu, g(ue)) = (gu, (gu)e)$ because of the right-associativity of the subset E.]

7.14 Interpret Exercise 7.13 in terms of Cayley digraphs. What is M in this case?

7.7 Notes and guide to references

The Petersen graph is a remarkable graph that appears in so many different graph-theoretical questions that it seemed worth dedicating a whole book to it. In fact, this has been done by Holton and Sheehan [108]. More information about generalised Petersen graphs, Kneser graphs, Kneser's Conjecture and odd graphs can be found in [108].

The Theorem of Erdös, Ko and Rado can be found in [67].

The proofs of Theorems 7.1 and 7.2 can be found in [80] and [154, 204], respectively.

A more detailed description of the Frucht notation can be found in the original paper [79], or in [51], pp. 211–216. Exercise 7.8 is from [57].

Theorem 7.5 is from [85].

The case $k = 2$ of Theorem 7.5 was first settled by Sabidussi [225].

A complete characterisation of metacirculant Cayley graphs can be found in [8], Theorem 12.

Exercise 7.12(a) is from [5, 225] while Exercise 7.12(b), which is from [248], confirms Ádám's Conjecture [2] for circulant groups of the cyclic group $\mathbb{Z}_p$. This conjecture states that $\text{Cay}(\mathbb{Z}_n, S)$ is isomorphic to $\text{Cay}(\mathbb{Z}_n, T)$ if and only if there exists a $z \in \mathbb{Z}_n$ such that $T = zS$. Although this conjecture has been shown to be false in general [66], obtaining those cyclic groups for which it holds attracted considerable interest (see, for example, [123, 197, 198]). A generalisation of this question asks: for which Cayley graphs $\text{Cay}(\Gamma, S)$, $\text{Cay}(\Gamma, T)$ is it true that $\text{Cay}(\Gamma, S) \simeq \text{Cay}(\Gamma, T)$ if and only if there is a group automorphism σ of Γ such that $S = \sigma(T)$? Exercise 5.8 shows that this is true for GRRs. More information on this problem can be found in [6].

Theorem 7.8 is from [63]. Theorem 7.7 is from [248].

The classification theorems for pq-graphs and for pq-digraphs have been obtained in [172] and [171], respectively. A basic step towards this result has been the solution of the imprimitive case, by detecting the peculiar case of MS-graphs ([169]) and then showing that any further imprimitive pq-graph is necessarily metacirculant [168]. The primitive case splits into several special classes, described in [167], [170] and [173]. A fuller review of the solution of this problem is given in [230].

8

The Reconstruction Conjectures

The Reconstruction Problem is considered one of the foremost unsolved problems in graph theory. In this chapter we shall define the two principal variants of this problem, and we shall give some of the most basic results related to these two problems. We shall also try to convey the flavour of what it takes to reconstruct a class of graphs and how this is related to important graph theoretic properties of the class.

A look at some of the surveys on the Reconstruction Problem, for example, [31, 32, 140, 203], will show that a considerable number of papers on this topic have been published. Also, many reconstruction proofs involve ad hoc case-by-case analysis; at this stage of our knowledge of the problem, it seems impossible to simplify these proofs significantly. It is therefore not the aim of the remaining chapters of this book to give anything like an exhaustive treatment of these results. The reader can see a more complete coverage of the Reconstruction Problem by consulting surveys such as the ones cited earlier and then going to the papers themselves where the results were obtained.

What we shall do in this chapter is give those basic definitions, techniques and results which can be presented succinctly and which are often used in most work on the reconstruction of graphs. We shall also discuss in some detail the reconstruction of particular classes of graphs. In the next chapter we shall look at variations of the main Reconstruction Problem that have been formulated, and in the final two chapters we shall present two general techniques that have been developed and that allow a systematic treatment of reconstruction rather than an ad hoc one for particular graph classes. In these four chapters we shall also try to emphasise as much as possible those aspects of the reconstruction of graphs that are related to symmetry properties of graphs, and we shall as much as possible try to present results that use ideas and theorems that we have already presented in earlier chapters.

G:

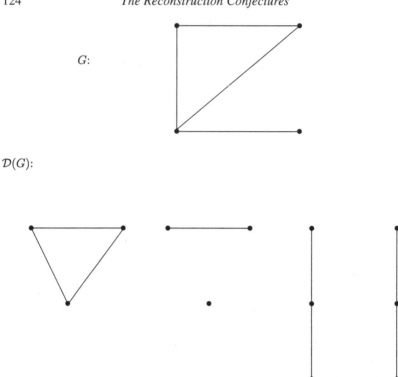

$\mathcal{D}(G)$:

Figure 8.1. A graph and its deck

8.1 Definitions

Given a graph G, the collection of its vertex-deleted subgraphs $G - v$ for all $v \in V(G)$ is called the *deck* of G and is denoted by $\mathcal{D}(G)$. Note that the graphs in the deck, which are sometimes referred to as *cards*, are unlabelled and, if G contains isomorphic vertex-deleted subgraphs, then such subgraphs are repeated in $\mathcal{D}(G)$ according to the number of isomorphic subgraphs that G contains. Therefore $\mathcal{D}(G)$ is a multiset, rather than a set, of isomorphism types of graphs. Figure 8.1 shows an example of a graph and its deck.

Suppose H is another graph with $\mathcal{D}(H) = \mathcal{D}(G)$. The question we are interested in is, 'Must H be isomorphic to G?' A graph H with $\mathcal{D}(H) = \mathcal{D}(G)$ is called a *reconstruction* of G. If every reconstruction of G is isomorphic to G, then G is said to be *reconstructible*. A graph that is not reconstructible is given by $G = K_2$ because, if H is the graph consisting of two isolated vertices, then clearly H is a reconstruction of G but it is not isomorphic to G. The Reconstruction Conjecture claims that these are the only non-reconstructible graphs.

The Reconstruction Conjecture *Every graph with at least three vertices is reconstructible.*

Another way of looking at reconstruction is by saying that a graph G is reconstructible if it can be uniquely (up to isomorphism) determined from $\mathcal{D}(G)$. Note that the problem stated in this way is not about finding an efficient algorithm for reconstructing G from $\mathcal{D}(G)$. In principle, given the deck, one can consider all graphs on n vertices to check which of them have the given deck. The question remains one of uniqueness, that is, whether this search will find only one graph with the given deck. The deck is a collection of isomorphism types with appropriate multiplicities, and the question is whether the isomorphism type of G can be determined uniquely from this collection.

Closely related to this problem is the Edge-Reconstruction Problem. The *edge-deck* of G is analogously defined as the collection of edge-deleted subgraphs $G - e$, called *edge-cards*, for all edges $e \in E(G)$; the edge-deck of G is denoted by $\mathcal{ED}(G)$. An *edge-reconstruction* of G is a graph H such that $\mathcal{ED}(H) = \mathcal{ED}(G)$, and G is said to be *edge-reconstructible* if every edge-reconstruction of G is isomorphic to it. The graph $G = K_3 \cup kK_1$ is not edge-reconstructible because, if H is the graph $K_{1,3} \cup (k - 1)K_1$, then H is an edge-reconstruction of G that is not isomorphic to it. Also, $G = 2K_2$ is not edge-reconstructible because if $H = P_3 \cup K_1$ (where P_3 is the path on three vertices), then $\mathcal{ED}(G) = \mathcal{ED}(H)$ but $G \not\cong H$.

The Edge-Reconstruction Conjecture claims that these are the only graphs that are not edge-reconstructible.

The Edge-Reconstruction Conjecture *Every graph on at least four edges is edge-reconstructible.*

Intuition seems to suggest that it is easier to reconstruct a graph from its edge-deck than from its deck; there are generally more graphs in the edge-deck, and edge-deleted subgraphs are generally more nearly like the original graph than vertex-deleted subgraphs. This intuitive notion will be borne out by Theorem 8.8, which essentially says that if the Reconstruction Conjecture is true, then so is the Edge-Reconstruction Conjecture.

It is well to emphasise here that at the heart of the difficulty of reconstructing a graph G from its deck is the symmetry of G and of the subgraphs in its deck. We shall give two illustrations of this. Let us first consider an extreme case. Suppose that the vertices of G are labelled $1, \ldots, n$ and that these labels are preserved on every vertex-deleted subgraph. Then, clearly, the graph G can be uniquely reconstructed by considering any three subgraphs in its deck and 'superimposing' them accordingly. This happens because the labellings have essentially removed all symmetries of G and its subgraphs, and therefore all ambiguities of how these subgraphs are embedded inside G. This situation is, in fact, shown to hold for most graphs in Lemma 9.3. Exercise 8.1 also shows that reconstruction follows easily if the symmetries of G and its subgraphs are suitably restricted. The final chapter of this book brings out more clearly the role of Aut(G) in the edge-reconstruction of G.

As a second illustration that the reconstruction is a question of graph symmetries, consider the problem this way. If G and H have the same deck, then we can define a bijective mapping ψ from $V(G)$ to $V(H)$ such that $G - v$ is isomorphic to $H - \psi(v)$. Such a mapping is often called a *hypomorphism* from G to V. The Reconstruction Conjecture now states that if there is a hypomorphism from G to H, then there is also an isomorphism between them. But to see that this cannot be immediately true notice that a hypomorphism can itself fail to be an isomorphism. For example, if both G and H are vertex-transitive, than all $G - v$ and $H - w$ are isomorphic, therefore any bijection from $V(G)$ to $V(H)$ is a hypomorhism, but of course, not all can be isomorphisms, unless G is the complete graph or its complement. Since a hypomorphism need not be an isomorphism, is there any strong reason why the existence of a hypomorphism should force the existence of an isomorphism?

Faced with the severe difficulty of cracking the Reconstruction Conjectures, most work carried out has been of a partial nature. Some basic graph parameters have been shown to be reconstructible and also some classes of graphs. A parameter $\theta = \theta(G)$ is said to be *reconstructible* if, for all reconstructions H of $G, \theta(H) = \theta(G)$. In other words, $\theta(G)$ is reconstructible if it can be determined uniquely from the deck of G. A class C of graphs is said to be *reconstructible* if every graph in C is reconstructible. Edge-reconstructible parameters and edge-reconstructible classes are analogously defined.

In the next sections we shall see some examples of reconstructible classes and parameters.

8.2 Some basic results

When considering the reconstruction (or, respectively, the edge-reconstruction) of a graph G we shall tacitly assume that the order of G is at least 3 (or, respectively, its size is at least 4).

Lemma 8.1 *Let G be a graph with n vertices and m edges. Then*

$$m = \frac{1}{n-2} \sum_{v \in V(G)} |E(G - v)|.$$

Therefore the number of edges of G is reconstructible.

Proof First note that n is clearly reconstructible; it is simply the number of graphs in the deck. Now, $\deg(v)$ is equal to $m - |E(G - v)|$. Therefore, by the Handshaking Lemma, $\sum_v (m - |E(G - v)|) = 2m$, and from this the foregoing formula follows. Because the right-hand side of this formula is obviously reconstructible, so is m. $\qquad\square$

Corollary 8.2 *Given a graph $G - v$ in the deck of G, the degree of v and the degrees of the neighbours of v in G are reconstructible.*

Proof The degree of v in G is simply $m - |E(G - v)|$, and this is reconstructible since m is. Therefore $\mathbf{d}$, the degree sequence of G in nondecreasing order, is reconstructible. Let $\mathbf{d}'$ be the degree sequence of $G - v$ but with the degree of v inserted in its correct position. The nonzero entries of the vector difference $\mathbf{d} - \mathbf{d}'$ occur in positions corresponding to neighbours of v in G, and their degrees can be read off from $\mathbf{d}$. $\qquad\square$

We can now give a simple proof that a specific class of graphs is reconstructible.

Theorem 8.3 *Regular graphs are reconstructible.*

Proof From the deck of G the degree sequence is reconstructible. Therefore, from $\mathcal{D}(G)$ it can be determined whether G is regular, and, if it is, its degree d is, of course, reconstructible. If G is regular, take any $G - v$ in the deck. The only way to reconstruct a regular graph of degree d from $G - v$ is to add a new vertex joining it to all the vertices of degree $d - 1$ in $G - v$. Hence G is uniquely reconstructible. $\qquad\square$

It is very important to note the strategy in this proof. We are only given the deck of G and not the fact that G is regular. The first step is therefore to recognise the regularity of G. This first step is called the recognition of the class. A class $\mathcal{C}$ of graphs is said to be *recognisable* if, for all graphs G in $\mathcal{C}$, any reconstruction of G must be in $\mathcal{C}$. The next step is to reconstruct G from $\mathcal{D}(G)$ and the extra information that G is in the class $\mathcal{C}$. This is called weak reconstruction. A class $\mathcal{C}$ of graphs is said to be *weakly reconstructible* if, for all graphs G in $\mathcal{C}$, any graph in $\mathcal{C}$ that is a reconstruction of G is isomorphic to G. Therefore in the proof we first showed that the class of regular graphs is recognisable. With this extra piece of information, we proceeded to reconstruct the regular G, that is, we showed that the class of regular graphs is weakly reconstructible.

The same strategy is employed in the next theorem. The short proof presented here is due to Manvel.

Theorem 8.4 *Disconnected graphs are reconstructible.*

Proof Let G be a disconnected graph. If G has isolated vertices, then its minimum degree is zero, and this fact can be determined from $\mathcal{D}(G)$. Otherwise, G is disconnected if and only if every $G - v$ is disconnected. Therefore the class of disconnected graphs is recognisable.

We have determined that the graph G to be reconstructed is disconnected. Amongst all the components of all the graphs in $\mathcal{D}(G)$, let C be one with a maximal number of vertices. Then C must be a component of G. Let v_0 be a vertex of C that is not a cutvertex (such a vertex must always exist). Consider all graphs in $\mathcal{D}(G)$ that have the least number of components isomorphic to C. Amongst these, let $G - v$ be the one with the largest number of components isomorphic to $C - v_0$. Then the only way to obtain G from $G - v$ is by replacing one component $C - v_0$ by C. $\square$

The following is one result that, although quite simple, has proved perhaps the most useful single tool in all investigations of the Reconstruction Problem. If G and F are graphs, then $\binom{G}{F}$ will denote the number of subgraphs of G isomorphic to F.

Theorem 8.5 (Kelly's Lemma) *Suppose G and F are graphs with $|V(F)| < |V(G)|$. Then*

$$(|V(G)| - |V(F)|)\binom{G}{F} = \sum_{v \in V(G)} \binom{G - v}{F}.$$

Therefore $\binom{G}{F}$ is reconstructible.

Proof The right-hand side of the formula counts all subgraphs isomorphic to F in $\mathcal{D}(G)$. But each such subgraph occurs exactly $|V(G)| - |V(F)|$ times; hence the equation follows.

Since the right-hand side of this equation is reconstructible and $|V(G)|$ and $|V(F)|$ are known, $\binom{G}{F}$ is reconstructible. $\square$

[*Remark:* Another way to prove the formula is to note that both sides of the equation count the number of pairs $(H, G - v)$ with $H \simeq F$, $G - v \in \mathcal{D}(G)$ and H a subgraph of $G - v$.]

The converse of Kelly's Theorem has been proved by Dulio and Pannone in [65]. Here we give a simpler proof of this converse which will then be used in the proof of Theorem 8.8.

Theorem 8.6 (Converse of Kelly's Lemma) *Suppose that all the values of $\binom{G}{F}$ are known for all graphs F with $|V(F)| < |V(G)|$. Then the deck $\mathcal{D} = \mathcal{D}(G)$ can be obtained from these values.*

Proof Let G have n vertices and m edges. Let $S = \{G_1, G_2, \ldots, G_N\}$ be the set of all nonisomorphic graphs on $n - 1$ vertices. Let

$$\mathcal{D}'_0 = \mathcal{D}'_0(G) = \sum_{G_i \in S} \binom{G}{G_i} G_i$$

(The use of the suffix 0 will be made clear later.) This notation means that the multiset $\mathcal{D}'_0$ contains all graphs G_i on $n - 1$ vertices and such that each G_i appears $\binom{G}{G_i}$ times in $\mathcal{D}'_0$

We note the following facts about $\mathcal{D}'_0$. Firstly, the multiset $\mathcal{D}'_0$ is known from the values $\binom{G}{F}$ given in the theorem. Secondly, if we can show that $\mathcal{D}(G)$ can be obtained from $\mathcal{D}'_0$, then the theorem is proved. Thirdly, the reason that $\mathcal{D}'_0$ is not $\mathcal{D}$ is that $\mathcal{D}'_0$ contains all subgraphs of G on $n - 1$ vertices whereas $\mathcal{D}$ contains only the induced subgraphs of G on $n - 1$ vertices. Therefore, if we can show that the noninduced subgraphs can be weeded out from $\mathcal{D}'_0$, then we are done. We shall now proceed to do this.

Let $\mathcal{D}_1$ be the multiset of graph isomorphism types contained in $\mathcal{D}'_0$ (together with multiplicities) such that these graphs are not subgraphs of any other graphs in $\mathcal{D}'_0$. These are clearly induced subgraphs of G and they are in the deck of G; yet they (and hence $\mathcal{D}_1$) need not be the whole deck of G. Represent $\mathcal{D}_1$ with the notation given earlier, that is, as a sum of isomorphism types of graphs with the coefficients representing multiplicities. Let $\mathcal{D}'_1$ be obtained from $\mathcal{D}_1$ by adding to this sum all the subgraph types obtained by deleting edges in all possible ways and numbers from the graphs in $\mathcal{D}_1$, taking care to determine the correct multiplicities of these subgraphs. Consider the difference between the two sums

$$\mathcal{D}'_0 - \mathcal{D}'_1.$$

Determine those graphs contained in this difference and such that these graphs are not subgraphs of any other graphs in the difference; these are clearly induced subgraphs of G, and hence in the deck of G. Add these subgraphs (with their multiplicities as they appear in the difference) to $\mathcal{D}_1$ to give $\mathcal{D}_2$.

Repeat this procedure (deleting edges from graphs in $\mathcal{D}_2$) to give $\mathcal{D}_2'$. If $\mathcal{D}_2' = \mathcal{D}_0'$ then $\mathcal{D}_2$ is the deck of G. Otherwise, continue as earlier from the difference $\mathcal{D}_0' - \mathcal{D}_2'$, and repeat until $\mathcal{D}_r' = \mathcal{D}_0'$. When that happens, $\mathcal{D}_r$ is the deck of G, $\mathcal{D}(G)$. □

Perhaps an example will make this proof clearer. Let G be a square with a diagonal edge. Then

$$\mathcal{D}_0' = 2K_3 + 8P_2 + 10(P_1 \cup N_1) + 4N_3$$

where K_k is the complete graph on k vertices, P_k is the path on k edges and N_k is the null graph on k vertices. Then,

$$\mathcal{D}_1 = 2K_3$$

$$\mathcal{D}_1' = 2K_3 + 6P_2 + 6(P_1 \cup N_1) + 2N_3$$
$$\mathcal{D}_2 = 2K_3 + 2P_2$$
$$\mathcal{D}_2' = 2K_3 + 8P_2 + 10(P_1 \cup N_1) + 4N_3.$$

And since $\mathcal{D}_2' = \mathcal{D}_0'$, $\mathcal{D}_2$ is the deck of G, which can easily be verified as correct.

As we have already noted, it is reasonable to expect that the the Edge-Reconstruction Problem is, in some sense, easier than the vertex problem. We shall now make this sense more precise. First note that the analogues for edge-reconstruction of Lemma 8.1 and Kelly's Lemma can just as easily be proved. We shall state, for later reference, Kelly's Lemma for edge-reconstruction, whose proof is exactly analogous to the vertex-reconstruction version.

Theorem 8.7 (Kelly's Lemma for edge-reconstruction) *Suppose G and H are graphs with $|E(F)| < |E(G)|$. Then*

$$(|E(G)| - |E(F)|)\binom{G}{F} = \sum_{e \in E(G)} \binom{G-e}{F}.$$

Therefore $\binom{G}{F}$ is edge-reconstructible.

We can now prove the following important result.

Theorem 8.8 (Greenwell) *Let G be a graph without isolated vertices. The deck of G is edge-reconstructible, that is, $\mathcal{D}(G)$ is uniquely determined from $\mathcal{ED}(G)$. Therefore, if G is reconstructible, then it is also edge-reconstructible.*

Proof Suppose G has n vertices and m edges. Let $S = \{G_1, G_2, \ldots, G_N\}$ be the set of all nonisomorphic graphs on $n - 1$ vertices and less than m edges. Since

G has no isolated vertices, every isomorphism type that appears in $\mathcal{D}(G)$ also appears at least once in S.

Let $T = \{G^1, G^2, \ldots, G^M\}$ be the set of nonisomorphic graphs on n vertices and $m - 1$ edges. Again, each isomorphism type appearing in $\mathcal{ED}(G)$ appears once in T.

Let $P = P_{ij}$ be the $N \times M$ matrix defined by $P_{ij} = \binom{G^j}{G_i}$. Note that no row of P is all zeros because each G_i has $n - 1$ vertices and fewer than m edges. Similarly, no column of P is all zeros.

Now, we shall determine

$$\mathcal{D}_0'(G) = \sum_{G_i \in S} \binom{G}{G_i} G_i$$

noting that each isomorphism type of a subgraph of G on $n-1$ vertices appears once in S and $\binom{G}{G_i}$ times in $\mathcal{D}_0'(G)$. But then we are done because from $\mathcal{D}_0'(G)$ we can, by proceeding as in the proof of the converse of Kelly's Lemma, obtain the deck $\mathcal{D}(G)$ in which only the number of times each *induced* subgraph of G on $n - 1$ vertices appears in G is counted.

The edge-deck of G can also be represented as

$$\mathcal{ED}(G) = \sum_{G^j \in T} \binom{G}{G^j} G^j.$$

Now let $\binom{\mathcal{ED}(G)}{G_i}$ be the number of times that G_i appears in $\mathcal{ED}(G)$. Note that if G_i is in the deck of G, then $\binom{\mathcal{ED}(G)}{G_i}$ is greater than 0 (again since every graph in $\mathcal{D}(G)$ has fewer than m edges because G has no isolated vertices). Suppose G_i has m_i ($< m$) edges (recall that each G_i has $n - 1$ vertices). Then

$$\binom{\mathcal{ED}(G)}{G_i} = \binom{G}{G_i} \cdot (m - m_i) = \binom{\mathcal{D}(G)}{G_i} \cdot (m - m_i)$$

because $\binom{G}{G_i} = \binom{\mathcal{D}(G)}{G_i}$, since G_i has $n - 1$ vertices. Therefore $\binom{\mathcal{D}(G)}{G_i}$ can be determined from $\binom{\mathcal{ED}(G)}{G_i}$.

But $\binom{\mathcal{ED}(G)}{G_i}$ depends on how many times G_i appears in each G^j and also the number of times that G^j appears in G. More precisely,

$$\binom{\mathcal{ED}(G)}{G_i} = \binom{G^1}{G_i}\binom{G}{G^1} + \ldots + \binom{G^M}{G_i}\binom{G}{G^M}.$$

Therefore, if a is the transpose of the vector

$$\left(\binom{G}{G^1}, \ldots, \binom{G}{G^M} \right)$$

and b is the transpose of the vector

$$\left(\left(\begin{matrix}\mathcal{ED}(G)\\ G_1\end{matrix}\right),\ldots,\left(\begin{matrix}\mathcal{ED}(G)\\ G_N\end{matrix}\right)\right),$$

then $b = Pa$. But P is known, the vector a is known from the given edge-deck, and therefore the vector b can be calculated. Hence, each $\left(\begin{smallmatrix}\mathcal{ED}(G)\\ G_i\end{smallmatrix}\right)$ can be calculated and, as described earlier, $\left(\begin{smallmatrix}\mathcal{D}(G)\\ G_i\end{smallmatrix}\right)$ can then be determined. ☐

Therefore, for graphs without isolated vertices, if the Reconstruction Conjecture is true, then so is the Edge-Reconstruction Conjecture. (We shall henceforth tacitly assume, unless otherwise stated, that any graph to be reconstructed has no isolated vertices.) So if we can prove a result for vertex reconstruction, say that a certain class of graphs is reconstructible, then this result would automatically hold for edge-reconstruction.

However, the Edge-Reconstruction Problem holds considerable independent interest because several results are known in this case which have not yet been proved for vertex-reconstruction, and some elegant proof techniques have been developed for edge-reconstruction. We shall consider some of these results in detail in a later chapter.

8.3 Maximal planar graphs

In this section we shall focus our attention on maximal planar graphs. It will not be possible to give the full proof that such graphs are reconstructible, since this is quite long and involves particular attention to special properties of maximal planar graphs. But we hope that the partial results presented here will give the reader a good idea of how the reconstruction of a nontrivial class of graphs has usually been achieved.

We shall first show that maximal planar graphs with minimum degree at least 4 are edge-reconstructible. Later we shall see that this is not the best possible reconstruction result for such graphs. However, the case we treat is short enough to present here in full detail and at the same time it is sufficiently nontrivial to bring out the flavour of reconstruction results in general and maximal planar graphs in particular.

As usual, the first task is to show that the class of graphs under consideration is edge-recognisable. This next result in fact gives that all planar graphs are edge-recognisable.

Theorem 8.9 (Fiorini) *A connected graph of order at least 7 and minimum degree at least 3 is planar if and only if every edge-deleted subgraph is planar.*

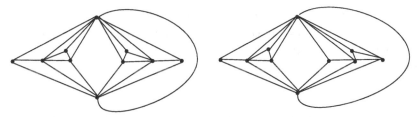

Figure 8.2. An example of two nonequivalent embeddings of $G - v$ where G is a maximal planar graph with $\deg(v) \geq 4$

Proof If a graph is planar, then clearly all edge-deleted subgraphs of the graph are planar. Therefore consider the converse. Suppose G is as in the theorem and all $G - e$ are planar. For contradiction, assume that G is nonplanar. Then G contains a subdivision of K_5 or $K_{3,3}$. Because G has at least seven vertices, then this subdivision must contain at least one vertex v of degree 2. But the degree of v in G is at least 3; therefore v is incident to an edge e that is not part of the subdivision and therefore $G - e$ is nonplanar, which is a contradiction. □

Corollary 8.10 *Maximal planar graphs of order at least 7 and minimum degree at least 3 are edge-recognisable.*

Proof Edge-recognition of planar graphs follows immediately from the theorem. Edge-recognition of maximal planar graphs follows because a planar graph is maximal if and only if it has $3n - 6$ edges, where n is the number of vertices, and the number of edges is reconstructible from the edge-deck. □

To proceed with the reconstruction we need the following results, which we give without proof. Note that a planar graph G is said to be *uniquely embeddable in the plane* if, for any two plane embeddings G_1, G_2 of G, there is an automorphism $\phi : G_1 \rightarrow G_2$ such that C is a cycle bounding a face in G_1 if and only if $\phi(C)$ is a cycle bounding a face in G_2. The best way to appreciate the meaning of unique embeddability in the plane is to consider an example of a graph which is not uniquely embeddable, as in Figure 8.2.

Theorem 8.11 (Whitney) *Every maximal planar graph is 3-connected. Also, a 3-connected planar graph is uniquely embeddable in the plane.*

Theorem 8.12 (Chartrand, Kaugars and Lick) *If G is a k-connected graph whose minimum degree is at least $(3k - 1)/2$, then there is a vertex v of G such that $G - v$ is still k-connected.*

Corollary 8.13 *If G is a maximal planar graph with minimum degree at least 4, then there is a vertex v of G such that G − v is 3-connected and therefore uniquely embeddable in the plane.*

Proof Follows immediately from the previous two theorems. ☐

We now have all the elements to prove our edge-reconstruction result.

Theorem 8.14 (Fiorini) *Maximal planar graphs with minimum degree at least 4 are edge-reconstructible.*

Proof Let G be a maximal planar graph with minimum degree at least 4. The first task is to show that the maximal planarity of G can be recognised from $\mathcal{ED}(G)$. But this is just Corollary 8.10.

Now, using Theorem 8.8, construct the vertex-deck $\mathcal{D}(G)$ of G. By Corollary 8.13 there is one $G−v$ in the deck that is 3-connected. Hence, by Whitney's Theorem, $G − v$ has a unique embedding in the plane. But knowing that G is maximal planar means that there is only one way to reconstruct from the plane embedding of $G − v$, namely, by joining a new vertex to all of the vertices on the boundary of the only nontriangular face of the embedding.

Hence G is uniquely determined from $\mathcal{ED}(G)$. ☐

At this point one might question why the *vertex-reconstruction* of maximal planar graphs with maximum degree 4 or 3 is not as easily carried out as in the proof of Theorem 8.14: just pick a vertex v with degree at least 4, embed $G − v$ in the plane, pick the face bounded by more than three edges, and join v to all the vertices bounding this face.

The first problem one faces is that of recognition, because it is not true, in general, that a graph G is planar if and only if every $G − v$ is planar, as the example in Figure 8.3 shows. However, Fiorini [73] has shown the following, relying heavily on Kuratowski's Theorem.

Theorem 8.15 (Fiorini) *Let G be a graph with minimum degree 5. Then G is planar if and only if every card G − v is planar.*

Therefore maximal planar graphs with mimimum degree 5 are recognisable (knowing that a maximal planar graph is a planar graph with $3n − 6$ vertices and knowing the number of edges of G from $\mathcal{D}(G)$). Using the theorem of Chartrand, Kaugars and Lick, it follows that such a maximal planar graph G has a vertex v such that $G − v$ is 3-connected and hence is uniquely embeddable

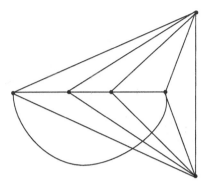

Figure 8.3. A graph which is not planar but each of whose cards is planar

in the plane leading to a unique reconstruction of G, as earlier. But more can be said. Fiorini and Manvel [76] proved the following for minimum degree 4.

Theorem 8.16 (Fiorini and Manvel) *Let G be a planar graph with minimum degree* 4. *Then, except for a finite number of graphs or families of graphs, all of which are reconstructible, G is planar if and only if every card $G - v$ is planar.*

Hence, even maximal planar graphs with minimum degree 4 are recognisable, although with considerably more effort, and therefore the argument can be extended to show that maximal planar graphs with minimum degree 4 are also reconstructible.

So, in order to finish off the reconstruction of maximal planar graphs we need consider only the case when the minimum degree is 3. Here, however, deciding which nonplanar graphs have the property that every card is planar becomes very complicated. But by restricting themselves to maximal planar graphs, Fiorini and Lauri [75] managed to show that maximal planar graphs with minimum degree 3 are recognisable.

Theorem 8.17 (Fiorini and Lauri) *A graph G on $n \geq 7$, m edges and at least two vertices of minimum degree* 3 *is maximal planar if and only if* (i) *$m = 3n - 6$ and* (ii) *each card $G - v$ is planar.*

We are now left with the task of reconstructing G knowing that it is maximal planar and has minimum degree 3. All we need is a vertex in G of degree at least 4 such that $G - v$ is 3-connected and therefore uniquely embeddable in the plane. However, it is easy to construct maximal planar graphs with minimum degree 3 in which no $G - v$ with $\deg(v) \geq 4$ is 4-connected. But in fact, in order to obtain reconstruction, all we need is that, at least for one vertex v with

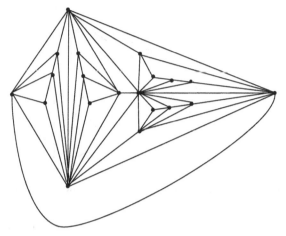

Figure 8.4. A maximal planar graph G such that all cards $G - v$ with $\deg(v) \geq 4$ have nonequivalent embeddings in the plane

degree at least 4, even if $G - v$ is not 3-connected, all its plane embeddings are equivalent. But, perhaps surprisingly, a maximal planar graph G can have the property that not only is $G - v$ not 3-connected for any vertex v of degree at least 4, but also that each such $G - v$ has at least two nonequivalent embeddings in the plane. A small example of such a maximal planar graph is shown in Figure 8.4.

To obtain the reconstruction of maximal planar graphs this was shown in [136]. Any such maximal planar graph G has at least one vertex v with $\deg(v) \geq 4$ such that (i) it has only two nonequivalent plane embeddings M_1, M_2 in which all faces are bounded by triangles except one bounded by the cycles C_1, C_2 in M_1, M_2, respectively, both of length $\deg(v)$, but (ii) the degrees of the vertices on C_1 are different from the degrees of the vertices on C_2. Since any maximal planar reconstruction is obtained by adding a new vertex and joining it to the vertices of C_1 or of C_2, knowledge of the degree sequence of G then determines which of the two embeddings is the right one to reconstruct from, giving the following.

Theorem 8.18 *A maximal planar graph is reconstructible.*

8.4 Digraphs and degree-associated reconstruction

The Reconstruction Conjecture is known to be false for digraphs. Harary and Palmer [102] demonstrated the first small counterexamples. Figure 8.5 shows a

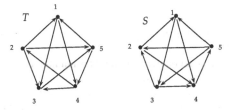

Figure 8.5. A pair of non-reconstructible tournaments on five vertices: $T_1 - i \simeq T_2 - i$ for $i = 1, \ldots, 5$ but $T_1 \not\simeq T_2$

pair of non-reconstructible tournaments on five vertices discovered by Beineke and Parker [23].

But the coup de grâce was delivered by Stockmeyer, who demonstrated in [237] the existence of infinite families of pairs of non-reconstructible tournaments. Each of these pairs of non-reconstructible tournaments is obtained by suitable joining together two tournaments which have the property that any vertex has a pseudosimilar mate. Full details are given in [32, 130, 237, 238].

Ramachandran, in [220], however, noted the following about these pairs of non-reconstructible tournaments. Let T_1, T_2 be such a pair. Then there are at least two corresponding cards $T_1 - v_1, T_2 - v_2$ of T_1 and T_2, respectively, such that although the cards are isomorphic, the in-degree/out-degree pairs of v_1 in T_1 and of v_2 in T_2 are not the same. Ramachandran therefore modified the Reconstruction Conjecture as follows for digraphs. Suppose G is a digraph. A *degree-associated card* of G is the triple $(G - v, \deg_{\text{in}}(v), \deg_{\text{out}}(v))$.

The Degree-Associated Reconstruction Conjecture for Digraphs *Every digraph with at least three vertices is reconstructible from its degree-associated cards.*

Ramachandran has shown that none of the known digraph counterexamples to the Reconstruction Conjecture are counterexamples to this Conjecture. A degree-associated card can also be defined for a graph G, when this time it would be the pair $(G - v, \deg(v))$. This gives no extra information when the full deck of G is available, but we shall see this notion of degree-associated reconstruction in action in the next chapter when we discuss reconstruction numbers. The analogous notion for edge-reconstruction is the degree-associated e-card which is the pair $(G - e, \deg(e))$ where $e = uv$ is an edge of G and $\deg(e) = \deg(u) + \deg(v) - 2$ which is therefore equal to the number of edges adjacent to e in G.

8.5 Exercises

8.1 Show that, if G has property A_2, then G is reconstructible and therefore almost every graph is reconstructible.

8.2 Give an example of a nonplanar graph with minimum degree at least 4 and at least seven vertices all of whose vertex-deleted subgraphs are planar.

8.3 A vertex v is said to be a good vertex if there is no vertex in G of degree equal to $\deg(v) - 1$. Show that if G has a vertex all of whose neighbours are good, then G is reconstructible.

8.4 Prove that, given any $G - e$, e an edge of G, then the degrees of the two vertices incident to e in G are reconstructible from $\mathcal{ED}(G)$. Show that a graph G with minimum degree δ is reconstructible in each of the following cases:

(a) G contains two adjacent δ-vertices;

(b) G contains a $(\delta + 1)$-vertex adjacent to two δ-vertices;

(c) G contains a triangle with one δ-vertex and two $(\delta + 1)$-vertices;

(d) G contains a triangle whose vertices all have degree $\delta + 1$, one of which is adjacent to a δ-vertex;

(e) G contains a $(\delta + k - 1)$-vertex adjacent to k δ-vertices.

8.5 A collection $S = \{H_1, H_2, ..., H_n\}$ is called an *illegitimate deck* if there is no graph on n vertices whose deck is equal to S. Let G be a graph of order n and let v_1, v_2 be two vertices such that $0 < \deg(v_1) - \deg(v_2) < n - 2$. Show that if, in the deck of G, $G - v_2$ is replaced by another copy of $G - v_1$, then the resulting collection of graphs is an illegitimate deck.

8.6 Show that a graph G is reconstructible if and only if its line graph $L(G)$ is edge-reconstructible.

8.7 Let G have a unique vertex v_0 of degree k and suppose that G contains no vertex of degree $k - 1$. Suppose also that no two vertices are removal-similar in $G - v_0$. Show that:

(a) the condition on the degree of v_0 is recognisable from $\mathcal{D}(G)$;

(b) the graph $G - v_0$ and the graphs $G - w$, for all neighbours w of v_0, are identifiable from $\mathcal{D}(G)$;

(c) if w is a neighbour of v_0, then the vertex v_0 is identifiable in $G - w$;

(d) given any $G - w$ with w neighbour of v_0, there is a unique vertex w' in $G - v_0$ such that $(G - v_0) - w' \simeq (G - w) - v_0$.

Deduce that the neighbours of v_0 in G are identifiable in $G - v_0$ and hence that G is reconstructible.

8.8 Some authors have considered the more difficult problem of reconstructing a graph G from its k-vertex-deleted subgraphs, that is, from its induced subgraphs on $n - k$ vertices (see, for example, [210, 211]). It is important, as in all reconstruction problems, that the vertices of the graph to be reconstructed are not implicitly labelled. Consider the following problems taken from Problems 14 and 15(a) in [153].

(a) Let G and H be two simple graphs sharing the same vertex-set V, $|V| > 3$, and suppose that $G - x - y$ and $H - x - y$ are isomorphic, for all pairs of vertices x, y. Show that $G = H$.

(b) However, show that if $G - x$ and $H - x$ are isomorphic for all $x \in V$, then G and H need not be identical.

8.6 Notes and guide to references

The reference [32] is a survey on the Reconstruction Problem giving extensive coverage of what was known up to 1977, but it is still a very useful reference for anyone wanting to learn about this problem for the first time. Our proof of Theorem 8.4 follows [162]. The proof we give of Theorem 8.8 is different from that in [91].

Motivated by Kelly's Theorem and its converse, Dulio and Pannone [65] proposed a variation of the Reconstruction Problem which can be described as follows. In the usual Reconstruction Problem, both G and H have the same number of subgraphs of all isomorphism types on at most $n - 1$ vertices. But instead Dulio and Pannone ask, would G and H be isomorphic if they contain the same number of subgraphs of some particular isomorphism type? For example, are two trees isomorphic if, for any caterpillar Q, they have an equal number of proper subgraphs isomorphic to Q? It is known that there can be two nonisomorphic trees having the same number of paths of any length.

Much more has been proved about the reconstruction of planar graphs and graphs embedded in other surfaces than what is presented in this chapter (see, for example, [70, 136, 266]). The proof we give of Fiorini's Theorem on the edge-reconstruction of maximal planar graphs is different from that in [74].

Whitney's Theorem was proved in [258] and Chartrand, Kaugars and Lick's Theorem in [54].

The results in Exercises 8.1 and 8.6 are from [55] and [92], respectively. The result in Exercise 8.1 will be substantially improved in Lemma 9.3 in the next chapter.

Relatively little is known about the illegitimate deck problem (again, see [32]). Deciding whether a given family of graphs is the deck of some graph is also an important problem from the point of view of issues of computational complexity, and it is closely related to the complexity of graph isomorphism (see [126, 134]).

Finally, note that the computer package [131] has a routine that can construct the deck of a given graph, and this could be useful when testing conjectures on small examples.

9

Reconstructing from Subdecks

Although graph theorists are still very far away from solving the Reconstruction Problem, in most cases where a class of graphs has been shown to be reconstructible it has turned out that only a few of the graphs in the deck were needed. So it seems that, in many cases at least, there might actually be more than sufficient information in $\mathcal{D}(G)$ to determine G uniquely. This has prompted many researchers to study variants of the Reconstruction Problem in which only some of the information in $\mathcal{D}(G)$ is given. Of course, these problems are more difficult than the original form of the Reconstruction Problem, so there is little hope of solving them in general. But their study has given rise to interesting problems, some of which use theory developed in earlier chapters. We shall consider in the next three sections three such variants of the Reconstruction Problem.

9.1 The endvertex-deck

One of the first classes of graphs that was shown to be reconstructible was trees [120]. This early result was subsequently improved so that it was shown that, for any tree T, the *endvertex-deck* $\mathcal{D}_1(T)$ consisting of those subgraphs $T - v$ with v an endvertex is sufficient to reconstruct T. One natural question that arises is therefore whether, given a graph G with a sufficiently large number of endvertices, G is *endvertex-reconstructible*, that is, reconstructible from its endvertex-deck $\mathcal{D}_1(G)$. Bryant has shown that this is, in fact, not true. His result is a clever use of Bouwer's Theorem and it requires the following lemma, whose easy proof is left as an exercise.

Lemma 9.1 *Let $F = \mathbb{Z}_2$ be the binary field and let X be the k-dimensional vector space consisting of all k-tuples of elements of F. Let Γ be the group of*

140

permutations that are linear transformations on X. Let A be a basis of X and let B be a set of k vectors of X whose sum is zero but such that any proper subset of B is linearly independent. Then,

  (i) *There is no $\alpha \in \Gamma$ such that $\alpha(A) = B$.*
  (ii) *If $A' \subset A$ and $B' \subset B$ each have $k - 1$ elements, then there is an $\alpha \in \Gamma$ such that $\alpha(A') = B'$.*

Theorem 9.2 (Bryant) *For any positive integer k there exists a graph G with k endvertices such that G is not endvertex-reconstructible.*

Proof Let Γ and X be as in the previous lemma, and let K be the graph obtained by means of Bouwer's Theorem, such that $X \subset V(K)$, $\mathrm{Aut}(K) \simeq \Gamma$, X is invariant under the action of $\mathrm{Aut}(K)$ and the action of $\mathrm{Aut}(K)$ on X is equivalent to that of Γ. As we have already observed, this construction can be carried out such that the minimum degree of K is at least 2.

Now let G be obtained from K by adding k new endvertices and joining each to a different vertex from A. Similarly, let H be obtained from K by adding k endvertices, one joined to each of the vertices in B. Then G is not endvertex-reconstructible because, by the previous lemma, G and H have the same endvertex-deck but $G \not\simeq H$. ☐

Remarks:

  (i) Note the similarity between this construction and that with which we produced graphs with all their endvertices mutually pseudosimilar in Section 5.7.
  (ii) Suppose K were any graph with degree at least 2, and let G be obtained from K by adding $|K|$ endvertices, joining each one to a different vertex of K. Clearly G would be endvertex-reconstructible. Therefore, although there can be graphs with an arbitrarily high number of endvertices that are not endvertex-reconstructible, it seems that there might be an upper bound on the *proportion* of endvertices in such graphs. In the next chapter we shall show that if the number of endvertices added to K is larger than $|K|/2$, then G would indeed be endvertex-reconstructible.

9.2 Reconstruction numbers

Since it seems that not all graphs in the deck might be necessary for reconstruction, one can ask for the minimum number of vertex-deleted subgraphs required to reconstruct G. This number is called the *reconstruction number* of

G, denoted by rn(G). (If G is a counterexample to the Reconstruction Conjecture, we let rn(G) = ∞. We know of no examples with this value of rn(G)!) A weakening of the problem of determining reconstruction numbers involves class-reconstruction numbers. Let C be a class of graphs and let $G \in C$. The *class-reconstruction number* of G, denoted by Crn(G), is defined as the least number of vertex-deleted subgraphs required to determine G given that $G \in C$. That is, Crn(G) is the least number of vertex-deleted subgraphs of G required to distinguish it from any other graph in C.

The *edge-reconstruction number* and the *class-edge-reconstruction number* are analogously defined.

Another reconstruction number which seems even more difficult to determine was defined by Myrvold [200]. The *adversary reconstruction number* of a graph G, denoted by adv-rn(G), is the minimum number k such that any k cards of G determine G uniquely. 'Adversary' analogues of the class- and the edge-reconstruction numbers can be defined. The reason for the name is that these two numbers can be seen as a game between two players, **A, B**. In the reconstruction number (also called the ally-reconstruction number by Myrvold) given a graph G, player **A** has to find the least number of cards which, given to the ally **B**, will allow the unique determination of G; this smallest number of cards is the reconstruction number of G. However, in the adversary reconstruction number, player **A** has to find the largest number h such that any h cards of G given to the adversary **B** will be insufficient to determine G uniquely; in this case, the adversary reconstruction number of G is equal to $1 + h$, that is, it is one more than the largest number of cards G has in common with any other graph H.

We have seen, in Exercise 8.1, that a graph with property A_2 is reconstructible, and so almost every graph is reconstructible. We shall now give a stronger result.

Lemma 9.3 (Müller; Myrvold; Bollobás) *Let G have property A_3. Then* rn(G) = adv-rn(G) = 3.

Proof Suppose G has property A_3, and let $u, v, w \in V(G)$. We shall show that G is uniquely reconstructible from just $G - u$, $G - v$ and $G - w$.

Note first that v is identifiable in $G - u$ and also u is identifiable in $G - v$; because, since G has property A_3 (and hence A_2), the only pair of vertices $x \in V(G - u)$, $y \in V(G - v)$ such that $G - u - x \simeq G - v - y$ are $x = v$ and $y = u$. Let $X = G - u - x$ and $Y = G - v - y$. There can only be one isomorphism from X to Y. For suppose α and β are two such isomorphisms.

Let $z \in V(X)$ such that $\alpha(z) \neq \beta(z)$. Then $X - z \simeq Y - \alpha(z) \simeq Y - \beta(z)$, contradicting property A_3. Therefore we can label X and Y uniquely, and, from $X = G - u$, we can determine uniquely all of the neighbours of v in G, except possibly u. All we need to know is whether u and v are adjacent. To determine this we repeat the procedure with $G - w$ instead of $G - u$. $\square$

From Theorem 2.7 and this lemma the following surprising corollary is immediate.

Corollary 9.4 *Almost every graph has reconstruction number equal to 3. In fact, almost every graph can be reconstructed from any three vertex-deleted subgraphs in its deck, that is, has adversary reconstruction number equal to 3.*

In recent years, another variant of the reconstruction number has attracted attention from researchers. We recall from the previous chapter Ramachandran's definition of a degree-associated card and a degree-associated e-card of G. This then gives rise to the *degree-associated reconstruction number* denoted by da-rn(G) and the *degree-associated edge-reconstruction number* denoted by da-ern(G), as well as the related degree-associated adversary reconstruction numbers. In all these variants of the reconstruction number, instead of cards or e-cards one is working with the degree-associated analogue.

We shall here limit ourselves to listing some results on reconstruction numbers and degree-associated reconstruction numbers in order to give the reader an idea of what has been done and which are some of the leading questions.

(i) [201, 192] The reconstruction number of a disconnected graph with at least two nonisomorphic components is 3. When all components are isomorphic on c vertices, the reconstruction number can be as large as $c + 2$.

(ii) [13] Let G be a disconnected graph all of whose components are isomorphic on c vertices. The only such graphs with rn $= c + 2$ are those all of whose components are isomorphic to the complete graph K_c. No such graphs have rn $= c + 1$. It is not known how large rn(G) can be in terms of c if its components are isomorphic to a graph H on c vertices with $H \not\simeq K_c$.

(iii) [193] The edge-reconstruction number of a disconnected graph with at least two nonisomorphic components is 2. When all components are isomorphic on c vertices, the reconstruction number can be as large as $c + 2$.

(iv) [14] Let G be a disconnected graph all of whose components are isomorphic with c edges. The only such graphs with ern $= c + 2$ are those all of whose components are isomorphic to $K_{1,c}$. No such graphs have rn $= c + 1$. It is not known how large ern(G) can be in terms of c if its components are isomorphic to a graph H on c edges with $H \not\simeq K_{1,c}$.

(v) [202] The reconstruction number of trees with at least five vertices is 3. Harary and Lauri [99] conjectured that the class reconstruction number of trees is at most 2. Welhan [256] has shown that this is true for trees without vertices of degree 2.

(vi) [98] The class-reconstruction number of maximal planar graphs is at most 2.

(vii) [15] Except for three known trees, the edge-reconstruction number of bicentroidal trees is 2. The authors also conjecture that, with only finitely many exceptions and except for three known infinite families, the edge-reconstruction number of a centroidal tree is also 2.

(viii) From the proof of Lemma 9.3 it also follows that the degree-associated reconstruction number of most graphs is at most 2.

(ix) Barrus and West [20] show that, with only one exception on six vertices, the degree-associated reconstruction number of a caterpillar is 2 unless the caterpillar is a star (in which case, da-rn $= 1$). They also conjecture that only finitely many trees have degree-associated reconstruction numbers equal to 3.

(x) [221] Since the degree-associated reconstruction number of $tK_{r,r}$ equals $r + 2$ for $t, r \geq 2$, then there is, for any k, a disconnected graph $G = tH$ with $H \neq K_c$ such that rn$(G) \geq$ da-rn$(G) = k$. This shows that, in answer to the question in (ii), the reconstruction number of such a disconnected G where H has c vertices can be as large as $O(\sqrt{c})$.

(xi) [12] If $G = tH$, all edge-cards of H are isomorphic and $\delta(H) \geq 3$, then the degree-associated edge-reconstruction number of G is 2. Asciak conjectures that the result would still hold if the condition $\delta(H) \geq 3$ is replaced by $H \neq K_{1,3}, K_{1,2}$ or $K_{2,3}$.

Also, Asciak shows that the degree-associated edge-reconstruction number of a caterpillar is at most 2 and he conjectures that the same holds for all trees.

Finally, results about the adversary reconstruction number can be found in [37, 38, 41, 200].

9.3 The characteristic polynomial deck

In the next chapter we shall see that the characteristic polynomial of a graph is reconstructible from its deck. In this section we want to look at a problem considered by Schwenk: Can the characteristic polynomial of the graph G be reconstructed if only the *characteristic polynomial deck* is given, that is, the characteristic polynomial of each $G - v \in \mathcal{D}(G)$? Schwenk has shown that nonisomorphic pairs of graphs exist with the same characteristic polynomial deck. Therefore this deck alone does not reconstruct G in general. However, the reconstructibility of the characteristic polynomial from the polynomial deck is still an open question. We shall therefore restrict ourselves to presenting a few results that lead to a condition which guarantees that the characteristic polynomial of G can be reconstructed from its polynomial deck.

In the sequel, $\phi(G; x)$ will denote the characteristc polynomial of G. An $n \times n$ diagonal matrix with terms $x_1, x_2, \ldots, x_n$ will be denoted by $\mathrm{diag}(x_1, x_2, \ldots, x_n)$.

For any nonzero vector x and any symmetric matrix T the expression xTx^t / xx^t will be called a *Raleigh quotient*. The following elementary result from linear algebra is left as an exercise.

Lemma 9.5 *Let the symmetric matrix T have eigenvalues $\lambda_1 \geq \lambda_2 \geq \ldots \geq \lambda_n$. Then, for any nonzero vector x,*

$$\lambda_1 \geq \frac{xTx^t}{xx^t} \geq \lambda_n.$$

Lemma 9.6 *Let G have vertices $v_1, v_2, \ldots, v_n$. Then the derivative of the characteristic polynomial is given by*

$$\phi'(G; x) = \sum_{i=1}^{n} \phi(G - v_i; x).$$

Proof Let A be the adjacency matrix of G. In $\phi(G; x) = \det(xI - A)$ replace the x by variables $z_1, z_2, \ldots, z_n$. Differentiate using the multivariable form of the chain rule:

$$\phi'(G; x) = \sum_{i=1}^{n} \frac{\partial}{\partial z_i} \det(\mathrm{diag}(z_1, \ldots, z_n) - A) \frac{dz_i}{dx}.$$

But every $z_i = x$, so the factor $\frac{dz_i}{dx} = 1$ and can be dropped. By expanding the determinant along row i we see that the partial derivative with respect to z_i evaluated at $z_1 = \ldots = z_p = x$ is simply $\phi(G - v_i; x)$. This completes the proof. $\square$

Therefore the characteristic polynomial deck gives the characteristic polynomial of G up to a constant.

Theorem 9.7 (The Interlacing Theorem) *Let A be an n × n symmetric matrix with eigenvalues*

$$\lambda_1 \geq \lambda_2 \geq \ldots \geq \lambda_n,$$

and let B be obtained from A by removing its i-th row and i-th column and suppose B has eigenvalues

$$\mu_1 \geq \mu_2 \geq \ldots \geq \mu_{n-1}.$$

Then the eigenvalues of B interlace those of A, that is,

$$\lambda_i \geq \mu_i \geq \lambda_{i+1}.$$

Proof Let N be the $n - 1 \times n$ matrix whose rows are the standard basis of $\mathbb{R}^n$ but with the i-th such vector missing. Then $B = NAN^t$.

Let $e_1, \ldots, e_n$ be an orthonormal basis of eigenvectors of A corresponding to $\lambda_1, \ldots, \lambda_n$, and let $f_1, \ldots, f_{n-1}$ be an orthonormal basis of eigenvectors of B corresponding to $\mu_1, \ldots, \mu_{n-1}$.

Now fix i and let U be the subspace spanned by the vectors $f_1, \ldots, f_i$. Then

$$\frac{xBx^t}{xx^t} \geq \mu_i$$

for any nonzero vectors x in U. Also, let $W = \{xN : x \in U\}$. Then W is still an i-dimensional space and

$$\frac{yAy^t}{yy^t} \geq \mu_i$$

for any y in W.

But now let X be the space spanned by $e_i, \ldots, e_n$. There must be a nonzero vector y in $W \cap X$, and for this vector we have

$$\frac{yAy^t}{yy^t} \leq \lambda_i.$$

Therefore $\lambda_i \geq \mu_i$.

By a similar argument, $\mu_i \geq \lambda_{i+1}$ is proved commencing with U as the subspace spanned by $f_i, \ldots, f_{n-1}$. $\square$

Corollary 9.8 *Suppose one of the characteristic polynomials $\phi(G - v; x)$ has a repeated root. Then $\phi(G; x)$ is reconstructible from the characteristic polynomial deck.*

Proof By the Interlacing Theorem, the repeated root is also a root of $\phi(G; x)$. But the characteristic polynomial deck gives $\phi(G; x)$ up to the constant, and knowing one root gives this constant. □

9.4 Exercises

9.1 Prove Lemma 9.1.

9.2 Let $T = \{1, 2, \ldots, 2k\}$ and let Γ be the alternating group acting on T. Let $X = T \times T$ and let Γ act on X in the obvious way, that is, $\alpha : (a, b) \mapsto (\alpha(a), \alpha(b))$. Let $A = \{(1, 2), (3, 4), \ldots, (2k - 1, 2k)\}$ and $B = A \cup \{(2, 1)\} - \{(1, 2)\}$. Prove that,

(a) $|A| = |B| = k$;
(b) if $A' \subset A$ and $B' \subset B$ have size $k - 1$, then $\alpha(A') = B'$ for some $\alpha \in \Gamma$;
(c) there is no $\alpha \in \Gamma$ such that $\alpha(A) = B$.

Hence deduce Bryant's Theorem.

9.3 Prove that $\mathrm{rn}(G) \geq 3$ for any graph G.

9.4 Find a disconnected graph G consisting of isomorphic components each of order c and such that $\mathrm{rn}(G) = c + 2$.

9.5 Let $\mathcal{C}$ be the class of maximal planar graphs and let $G \in \mathcal{C}$ have minimum degree at least 4. Show that $\mathcal{C}\mathrm{rn}(G) = 1$.

9.6 Let $\mathcal{C}$ be the class of trees. Which trees T have $\mathcal{C}\mathrm{rn}(T) = 1$?

9.7 A graph is said to have *property* EA_k if $G - A \not\cong G - B$ for any two distinct subsets A, B of $E(G)$ with $|A| = |B| = k$. Prove that,

(a) if G has property A_{13}, then it has property EA_3;
(b) if G has property EA_3, then it is reconstructible from any two edge-deleted subgraphs in its edge-deck;
(c) almost every graph has edge-reconstruction number equal to 2.

9.8 Prove Lemma 9.5.

9.9 Let

$$\phi(G; x) = x^n + a_1 x^{n-1} + a_2 x^{n-2} + \ldots + a_n.$$

Show that $-a_2$ equals the number of edges of G. Deduce that the characteristic polynomial of a regular graph is reconstructible from its polynomial deck.

9.5 Notes and guide to references

Bryant's actual construction of the permutation group with the properties required for Theorem 9.2 is given in Exercise 9.2.

More is known about reconstruction numbers of certain classes of graphs than is presented here; see, for example, [19, 98, 99, 202]. An intriguing question that hardly has been given any consideration is the relationship between $\mathrm{rn}(G)$ and $\mathrm{ern}(G)$ for some given graph G. While, as we have seen, the truth of the Vertex Reconstruction Conjecture would imply the Edge-Reconstruction Conjecture, no similar simple relationship between reconstruction numbers and edge-reconstruction numbers seems to hold.

Reconstruction numbers are also interesting from another point of view. We have commented elsewhere that a graph without symmetries is easier to reconstruct (see, in fact, Exercise 8.1). However, a vertex-transitive graph is regular, and we have seen in the previous chapter that reconstructing regular graphs is almost trivial. But reconstruction numbers put this issue in a possibly better perspective, because those graphs with highest reconstruction number seem, in fact, to be regular graphs [200]. Indeed, the reconstruction number of such graphs is not yet known, whereas, as we have seen in this chapter, any three subgraphs from the deck of a graph with property A_3 would be sufficient to reconstruct the graph uniquely.

The problem of reconstructing the characteristic polynomial from the polynomial deck was posed in [231], from which the proof of Lemma 9.6 is taken. The proof we give of the Interlacing Theorem is based on [252]. Recently Cvetković and Lepović [59] have shown that the characteristic polynomial of a tree is reconstructible from its characteristic polynomial deck. Interestingly also from the point of view of later results that we shall present on endvertex-deck reconstruction, in [232] it is proved, amongst other things, that if a graph of order n has at least $n/3$ endvertices, then its characteristic polynomial is also reconstructible from the characteristic polynomial deck. In a very different spirit are references [89, 265] which give conditions on the eigenvalues and eigenvectors of G that ensure that it is reconstructible.

Finally note that another way of reconstructing by not using all the information in the deck, and which has received some attention, is the reconstruction of G from its *set* of vertex-deleted subgraphs, that is, repetitions of isomorphic graphs are removed from $\mathcal{D}(G)$. For more information on this problem see [161, 162].

10

Counting Arguments in Vertex-Reconstruction

Most results in reconstruction have been obtained by specific consideration of the particular class of graphs being studied. However, a number of general results have been obtained, and these involve very clever counting arguments. We shall look at these results in this and the following chapter.

10.1 Kocay's Lemma

Let G be a graph and $\mathcal{F} = (F_1, F_2, \ldots, F_k)$ a sequence of graphs (we do not exclude the possibility that different F_i could be isomorphic). A *cover* of G by $\mathcal{F}$ is a sequence $\mathcal{G} = (G_1, G_2, \ldots, G_k)$ of subgraphs of G (not necessarily distinct) such that

(i) $G_i \simeq F_i$, $i = 1, \ldots, k$;

(ii) $G = \cup_i G_i$.

The number of covers of G by $\mathcal{F}$ is denoted by $c(\mathcal{F}, G)$.

We collect as a lemma a few simple but important observations. Their easy proof is left as an exercise.

Lemma 10.1 *Let G be a graph and $\mathcal{F} = (F_1, F_2, \ldots, F_k)$ a sequence of graphs. Then,*

(i) *if each $F_i = K_2$ and $k = m = |E(G)|$, then $c(\mathcal{F}, G) = m!$;*

(ii) *if each $F_i = K_2$ and $k > m$, then $c(\mathcal{F}, G)$ is equal to the number of surjections from a k-set to an m-set that is equal to $m! S(k, m)$ (where $S(k, m)$ are the Stirling numbers of the second kind);*

(iii) *if $|E(G)| > \sum |E(F_i)|$, then $c(\mathcal{F}, G) = 0$; similarly if $|V(G)| > \sum |V(F_i)|$;*

(iv) *suppose that $\sum |V(F_i)| = |V(G)|$ and $c(\mathcal{F}, G) > 0$; then G is disconnected with k components isomorphic to $F_1, \ldots, F_k$;*

(v) *suppose that $\sum(|V(F_i)| - 1) = |V(G)| - 1$ and $c(\mathcal{F}, G) > 0$; then either G is disconnected or else it is separable with blocks isomorphic to $F_1, \ldots, F_k$;*

(vi) *suppose that G is disconnected (separable) with k components (blocks) isomorphic to $F_1, \ldots, F_k$ and that there are $s \leq k$ distinct isomorphism types of components (blocks), and suppose that G contains a_i, $1 \leq i \leq s$, components of the i-th isomorphism type; then*

$$c(\mathcal{F}, G) = \prod_{i=1}^{s} a_i!.$$

Theorem 10.2 (Kocay's Lemma) *Let G be a graph and let*

$$\mathcal{F} = (F_1, \ldots, F_k)$$

be a sequence of graphs with each $|V(F_i)|$ less than $|V(G)|$. Then

$$\prod_{i=1}^{k} \binom{G}{F_i} = \sum_{X} c(\mathcal{F}, X) \binom{G}{X}, \qquad (10.1)$$

where the sum extends over all isomorphism types of graphs. Therefore the parameter

$$\kappa(\mathcal{F}, G) = \sum_{X:|V(X)|=|V(G)|} c(\mathcal{F}, X) \binom{G}{X}$$

is reconstructible.

[*Remark:* Note that the sums in this theorem are not infinite because $\binom{G}{X} = 0$ if $|V(X)| > |V(G)|$ or $|E(X)| > |E(G)|$, and also $c(\mathcal{F}, X) = 0$ if $|V(X)| > \sum |V(F_i)|$ or $|E(X)| > \sum |E(F_i)|$.]

Proof Let S be the set of all pairs $(\mathcal{G}, H)$, where H is a subgraph of G and $\mathcal{G}$ is a cover of H by $\mathcal{F}$. We shall count $|S|$ in two ways.

Fixing any subgraph H of G gives $c(\mathcal{F}, H)$ elements of S. Therefore, all subgraphs of G isomorphic to H give, between them, $c(\mathcal{F}, H)\binom{G}{H}$ pairs. Doing this for all isomorphism types of subgraphs of G gives that $|S|$ equals the right-hand side of Equation (10.1).

Also, consider any sequence $\mathcal{G} = (G_1, \ldots, G_k)$ of subgraphs of G, $G_i \simeq F_i$ ($\mathcal{G}$ not necessarily a cover of G). Certainly, if $H = G_1 \cup \cdots \cup G_k$, then the pair

$(\mathcal{G}, H)$ belongs to S. Moreover, H is the only subgraph of G that is covered by this particular sequence $\mathcal{G}$. That is, a particular sequence $\mathcal{G}$ of subgraphs of G corresponds to one and only one H, that is, $H = G_1 \cup G_2 \ldots \cup G_k$. Therefore the number of pairs in S is equal to the number of such sequences $\mathcal{G}$, and this equals the left-hand side of Equation (10.1). This verifies the equation.

Now, the summation on the right-hand side of (10.1) is reconstructible because the left-hand side is reconstructible by Kelly's Lemma. But now consider partitioning this summation into three parts: $\sum_1$ containing all terms with $|V(X)| < |V(G)|$, $\sum_2$ containing terms with $|V(X)| = |V(G)|$, and $\sum_3$ containing the other terms with $|V(X)| > |V(G)|$. But $\sum_3 = 0$, $\sum_1$ is reconstructible by Kelly's Lemma, and therefore $\sum_2$ is reconstructible, and this is $\kappa(\mathcal{F}, G)$. $\qquad\qquad\square$

Kelly's Lemma from Chapter 8 says that from $\mathcal{D}(G)$ one can count all nonspanning subgraphs of G of a given isomorphism type. Of course, if one could count all spanning subgraphs of G of any given isomorphism type, then the Reconstruction Conjecture would be proved. But by a judicious choice of $\mathcal{F}$, Kocay's Lemma can be used to count certain types of spanning subgraphs of G. We shall see how this is done in the next section.

10.2 Counting spanning subgraphs

Theorem 10.3 (Tutte) *Let G be a graph and let*

$$\mathcal{F} = (F_1, \ldots, F_k),$$

$k \geq 2$, be a sequence of graphs with each $|V(F_i)|$ less than $|V(G)|$. Then the following parameters are reconstructible:

(i) *the number of disconnected spanning subgraphs of G with k components isomorphic to $F_1, \ldots, F_k$;*

(ii) *the number of separable spanning subgraphs of G with k blocks isomorphic to $F_1, \ldots, F_k$.*

Proof (i) If $|V(G)| \neq \sum |V(F_i)|$, then the number of disconnected spanning subgraphs of G with components isomorphic to $F_1, \ldots, F_k$ equals zero. Therefore assume $|V(G)| = \sum |V(F_i)|$. Now $\kappa(\mathcal{F}, G)$ is reconstructible. In the expression for $\kappa(\mathcal{F}, G)$ we need only consider those X such that $c(\mathcal{F}, X) > 0$. By Lemma 10.1(iv), X must be the disconnected graph with k components isomorphic to $F_1, \ldots, F_k$, and there is therefore only one term $c(\mathcal{F}, X)\binom{G}{X}$ corresponding to this X in the summation for $\kappa(\mathcal{F}, G)$. But $c(\mathcal{F}, X)$ is known because $\mathcal{F}$ and X are known (its value is given in Lemma 10.1(vi)). Therefore $\binom{G}{X}$ is reconstructible.

(ii) In this case, if $|V(G)| - 1 \neq \sum(|V(F_i)| - 1)$, then the number of separable spanning subgraphs of G with blocks isomorphic to $F_1, \ldots, F_k$ equals zero. Therefore assume $|V(G)| - 1 = \sum(|V(F_i)| - 1)$. Again, $\kappa(\mathcal{F}, G)$ is reconstructible, and again, in the expression for $\kappa(\mathcal{F}, G)$ we need only consider those X such that $c(\mathcal{F}, X) > 0$. By Lemma 10.1(v), X must be either a disconnected graph or a separable graph with blocks isomorphic to $F_1, F_2, \ldots, F_k$. Therefore we can partition the summation in $\kappa(\mathcal{F}, G)$ into two parts, one involving the term with X disconnected and another involving a sum of terms with X separable. The first term is reconstructible by the first part of the theorem. We are therefore left with

$$\sum_Y c(\mathcal{F}, Y) \binom{G}{Y},$$

where the summation is taken over all separable spanning subgraphs of G with k components isomorphic to $F_1, \ldots, F_k$, and we therefore know that this sum is reconstructible. But $c(\mathcal{F}, Y)$ takes the same value for all such Y, and this value is known (Lemma 10.1(vi)). Therefore

$$\sum_Y \binom{G}{Y}$$

is reconstructible, and this is precisely the number of separable spanning subgraphs of G with k blocks isomorphic to $F_1, \ldots, F_k$.   □

Corollary 10.4 *Let G be a graph. Then the following parameters are reconstructible:*

(i) *the number of 1-factors of G;*
(ii) *the number of spanning trees of G.*

Proof First consider (i). If $n = |V(G)|$ is odd, then clearly the number of 1-factors of G is zero. If n is even, then the result follows immediately from Theorem 10.3(i) by letting each $F_i \simeq K_2$, the complete subgraph on two vertices, and $k = n/2$.

Similarly, (ii) follows from (ii) of Theorem 10.3 by letting each $F_i \simeq K_2$ and $k = n - 1$.   □

The previous theorem and its corollary dealt with disconnected or separable spanning subgraphs of G. Using very much the same ideas we can tackle some 2-connected spanning subgraphs.

Theorem 10.5 *Let G be a graph. Then the following parameters are recon-structible:*

(i) *the number of Hamiltonian cycles of G;*
(ii) *the number of 2-connected spanning subgraphs of G with a fixed num-ber h of edges.*

Proof We shall again be using the reconstructibility of the parameter $\kappa(\mathcal{F}, G)$, where the sequence $\mathcal{F} = (F_1, F_2, \ldots, F_k)$ is such that each $F_i \simeq K_2$.

We first tackle (i). Let $k = n$. Consider $\kappa(\mathcal{F}, G)$, which is reconstructible. Consider any nonzero term in the summation in $\kappa(\mathcal{F}, G)$. For such a term, since $c(\mathcal{F}, X) > 0$, each X is a graph on n vertices and at most n edges. Therefore X is either an n-cycle C_n or it is a disconnected or separable graph. Therefore the summation can be partitioned into three parts,

$$\sum_1 + \sum_2 + \sum_3 = \sum_X c(\mathcal{F}, X) \binom{G}{X} + \sum_Y c(\mathcal{F}, Y) \binom{G}{Y} + c(\mathcal{F}, C_n) \binom{G}{C_n},$$

where X ranges over all isomorphism types of disconnected graphs on n ver-tices and Y ranges over all isomorphism types of trees on n vertices. But the first two summations are reconstructible by the previous results, and hence so is $c(\mathcal{F}, C_n) \binom{G}{C_n} = n! \binom{G}{C_n}$. Therefore $\binom{G}{C_n}$, the number of Hamiltonian cycles in G, is reconstructible.

Next we come to (ii). First note that (i) is, in fact, a special case of (ii) with $h = n$ because there is only one isomorphism class of 2-connected graphs on n vertices and n edges: a cycle. However, when $h > n$, there is more than one isomorphism class of 2-connected graphs on h edges and n vertices. Let s_h be the number of 2-connected spanning subgraphs of G with h edges; (ii) tells us that s_h can be reconstructed, although we might not know how many of these subgraphs of G correspond to particular individual isomorphism classes.

Let k be any number between n and h. Again, let $\mathcal{F} = (F_1, F_2, \ldots, F_k)$ with each $F_i \simeq K_2$. The number $\kappa(\mathcal{F}, G)$ is reconstructible. Consider the nonzero terms in $\kappa(\mathcal{F}, G)$ and argue as earlier, namely, that nonzero terms in the sum-mation in $\kappa(\mathcal{F}, G)$ corresponding to separable or disconnected X are recon-structible. This leaves the summmation

$$\sum_X c(\mathcal{F}, X) \binom{G}{X},$$

where X ranges over all isomorphism types of 2-connected graphs on n ver-tices; this summation is therefore reconstructible. However, for any such X with j edges, $c(\mathcal{F}, X)$ is equal to $j! S(h, j)$, the number of surjections from a

k-set to a j-set. Therefore, by grouping together in this summation all isomorphism types of 2-connected X with the same number of edges, we can rewrite it as

$$\sum_{j=n}^{k} j! S(k,j) s_j.$$

This can be repeated for all values of k such that $n \le k \le h$, giving $h - n + 1$ equations in the unknowns $s_n, s_{n+1}, \ldots, s_h$, which can therefore be determined. $\square$

10.3 The characteristic and the chromatic polynomials

The most important applications of Tutte's Theorems are in the reconstruction of the characteristic and the chromatic polynomials. We first need results that tell us how these two polynomials can be written in terms of certain types of subgraphs.

An *elementary graph* is a graph in which every component is either an edge or a cycle. For any graph X, $c(X)$ denotes the number of components of X and $s(X)$ the number of cycles. The proof of the following theorem of Sachs can be found as Proposition 7.3 in Biggs' book [24].

Theorem 10.6 (Sachs) *Let the characteristic polynomial of G be*

$$\lambda^n + a_1 \lambda^{n-1} + a_2 \lambda^{n-2} + \ldots + a_n.$$

Then each coefficient a_i is given by

$$a_i = \sum_{X} (-1)^{c(X)} 2^{s(X)} \binom{G}{X},$$

where the summation extends over all isomorphism types X of elementary graphs on i vertices.

The following theorem of Whitney can also be found as Theorem 10.4 in [24].

Theorem 10.7 (Whitney) *Let the chromatic polynomial of G be*

$$b_1 x + b_2 x^2 + \ldots + b_n x^n.$$

Then each coefficient b_i is given by

$$b_i = \sum_X (-1)^{|E(X)|} \binom{G}{X},$$

where the summation extends over all isomorphism types X of graphs on n vertices and i components.

We can now prove the main theorem of this section.

Theorem 10.8 (Tutte) *The characteristic polynomial and the chromatic polynomial are reconstructible.*

Proof Consider the characteristic polynomial as given by Theorem 10.6. For $0 \le i < n$, each a_i is reconstructible by Kelly's Lemma. For $i = n$, each elementary X appearing in the expansion for a_n is either disconnected or a Hamiltonian cycle. Therefore, by Corollary 10.4, each $\binom{G}{X}$, and hence a_n, is reconstructible.

Now consider the chromatic polynomial as given by Theorem 10.7. For $1 < i \le n$, each X appearing in the expansion for b_i is a disconnected spanning subgraph of G. Therefore b_i is reconstructible by Corollary 10.4. When $i = 1$, each X appearing in the expansion of b_1 has one component, that is, it is connected. The expansion for b_1 can be written as

$$b_1 = \sum_k (-1)^k c_k,$$

where each c_k is the total number of connected subgraphs of G with k edges, whether separable or 2-connected. This sum can be reconstructed by Theorems 10.3 and 10.5. □

10.4 Exercises

10.1 Prove Lemma 10.1.

10.2 Using (i) of Theorem 10.3, give another proof that disconnected graphs are reconstructible.

10.3 Let G be a graph on n vertices and $m \ge n$ edges. Let C_t denote the cycle on t vertices. Prove that

$$\binom{G}{C_n} = \binom{m}{n} - \sum_{t=3}^{n-1} \binom{G}{C_t} \cdot \binom{m-t}{n \quad 3} + \sum_X b_X \cdot \binom{G}{X},$$

where the last summation ranges over all isomorphism types X of graphs with n edges, no isolated vertices and containing at least two cycles, and where $b_X =$ (number of cycles in X) $-$ 1. Deduce that $\binom{G}{C_n}$ is reconstructible.

10.4 Look up the proofs of Theorem 10.6 and Theorem 10.7 on the characteristic and chromatic polynomials.

10.5 Notes and guide to references

Tutte's Theorems first appeared in [251]. His methods were greatly simplified by Kocay in [127, 128]. The presentation given in this chapter of Kocay's methods is based on [31].

11

Counting Arguments in Edge-Reconstruction

Perhaps the most significant results obtained on the Reconstruction Problem are the methods of Lovász and Nash-Williams in edge-reconstruction. We shall discuss here these methods in a slightly more general setting than the edge-reconstruction of graphs.

11.1 Definitions and notation

Let D (domain) be a finite set and let Γ be a group of permutations of D. A *structure* S is a triple $(D, \Gamma, \mathcal{E})$, where $\mathcal{E} = \mathcal{E}(S)$ is a subset of D.

If S_1, S_2 are two structures on the same D and Γ, then they are said to be *compatible*. Two compatible structures $S_1 = (D, \Gamma, \mathcal{E}_1)$ and $S_2 = (D, \Gamma, \mathcal{E}_2)$ are said to be *isomorphic under* Γ if there exists an $\alpha \in \Gamma$ such that for all $x \in D$, $x \in \mathcal{E}_1$ if and only if $\alpha(x) \in \mathcal{E}_2$. In this case we write $S_1 \simeq S_2$ or $S_1 \simeq_\Gamma S_2$ if we need to emphasise the role of Γ.

Examples

1) Let D be the set of cells in a 3×3 square grid, and let Γ be the dihedral group of order 8 acting on the square (that is, the group of all symmetries of the square). Let $S_1 = (D, \Gamma, \mathcal{E}_1)$ be the structure represented by

where the cells in $\mathcal{E}_1$ are those denoted by •.

157

Similarly, let $S_2 = (D, \Gamma, \mathcal{E}_2)$ be

and let $S_3 = (D, \Gamma, \mathcal{E}_3)$ be

Then S_1 and S_2 are isomorphic under Γ, but they are not isomorphic to S_3.
2) Let S_1 and S_2 be as in the previous example, but this time let Γ be the cyclic group of order 4 acting on the square, that is, the group of rotational symmetries of the square. Now S_1 and S_2 are not isomorphic under Γ.
3) Let $Y = \{1, 2, \ldots, n\}$ and let D be $\binom{Y}{2}$ the set of all unordered pairs of distinct elements of Y, that is, $D = \{\{x, y\} : x, y \in Y, x \neq y\}$. Let $\Gamma = S_n^{(2)}$, the group of permutations induced by S_n on the set of pairs D.

Of course, any such structure is simply a graph. Two graphs are isomorphic under Γ in the sense defined here if and only if they are isomorphic as graphs in the usual sense.

In this chapter, when a graph is considered as a structure it will be taken to mean as in this example.

This last example shows that the definition we have given for a structure is a simple extension of the definition of a graph. The following terminology and notation are motivated by this.

Given a structure S, the elements in $\mathcal{E}(S)$ are called the *edges* of S. Let $A \subseteq \mathcal{E}(S)$. Then the structure R whose edge-set is $\mathcal{E}(S) - A$ is called a *substructure* of S; R will be denoted by $S - A$. If $A = \{x\}$, then R will be written as $S - x$.

The number of substructures of S isomorphic to R is denoted by $\binom{S}{R}$.

If $B \subset D - \mathcal{E}(S)$, then the structure T whose edge-set is $\mathcal{E}(S) \cup B$ is said to be a *superstructure* of S. It is denoted by $S + B$, or $S + x$ when B contains just one element x.

The $\mathcal{S}$-*deck* of S, $\mathcal{SD}(S)$, is the collection of structures $S - x$ for all $x \in \mathcal{E}(S)$. Note that, in the $\mathcal{S}$-deck, the elements of D in each structure are unlabelled; that is, as in the decks of the graph reconstruction problems, the $\mathcal{S}$-deck really contains isomorphism types (under Γ) of structures with appropriate multiplicities.

A structure $S = (D, \Gamma, \mathcal{E})$ is said to be $\mathcal{S}$-*reconstructible* if it is true that, for any structure $T = (D, \Gamma, \mathcal{E}')$ with $\mathcal{SD}(T) = \mathcal{SD}(S)$, it follows that $T \simeq_\Gamma S$.

That is, S is S-reconstructible if it can be uniquely (up to isomorphism under Γ) determined from its S-deck or, rather, if its isomorphism type can be uniquely determined from the isomorphism types in the S-deck.

Note that when the structure S is as in Example 3, that is, a graph, then the S-deck of S is simply the edge-deck of the corresponding graph. Therefore the reconstruction of structures as given here is a slight extension of the Edge-Reconstruction Problem.

One advantage of looking at the problem in this more general setting is that certain variations that we have seen of the reconstruction theme, such as the problem of endvertex-reconstruction, can be described as special cases of the reconstruction of structures (see Example 5 and Exercise 11.9). Therefore the counting arguments that we shall be presenting here have wider applicability, apart from being clearer in this general setting. Also, this treatment brings out clearly what has already been mentioned in Chapter 8, namely, that at the heart of the problem of reconstruction is the group of symmetries acting on the structure to be reconstructed and on the substructures in its deck. Moreover, the availability of nontrivial structures that are not S-reconstructible (which we do not have for the Edge-Reconstruction Problem) places some of the ideas involved in better focus.

Finally, viewing the Edge-Reconstruction problem in this light has enabled a number of authors to extend the problem in various diverse directions, some of which we shall be mentioning briefly at the end of this chapter.

Examples

4) The structures S_1 and S_3 in Example 1 are simple examples of non-S-reconstructible structures because they have the same S-deck but they are not isomorphic.

5) Let H be a graph with minimum degree at least 2 and $\Gamma = \text{Aut}(G)$. Let $D = V(H)$. A structure $S = (D, \Gamma, \mathcal{E})$ can be seen as a graph G obtained from H by attaching endvertices to some of its vertices, precisely to those vertices in $\mathcal{E}$. The reconstructibility of S in this case is equivalent to the endvertex-reconstructibility of G.

11.2 Homomorphisms of structures

Let $R = (D, \Gamma, \mathcal{E}_1)$ and $T = (D, \Gamma, \mathcal{E}_2)$ be two structures. An *embedding* or *homomorphism* of R into T is a permutation $\alpha \in \Gamma$ such that, if $x \in \mathcal{E}_1$, then $\alpha(x) \in \mathcal{E}_2$. The set of homomorphisms from R to T is denoted by $(R \to T)$,

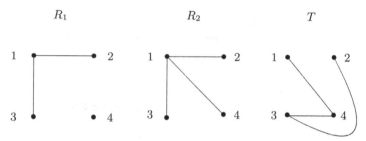

Figure 11.1. Illustrating embeddings of structures

and $[T]_R$ denotes the number of homomorphisms from the structure R to the structure T.

Examples

6) Let R be the structure S_1 of Example 1, and let T be the structure

.

   Suppose the cells in the grids are numbered 1 to 9 from left to right, starting at the first row. Then the permutation $\alpha = (2\ 4)(3\ 7)(6\ 8)$ is an embedding of R into T. However, if R were the structure

or the structure

   then there would have been no embedding of R into T.

7) Let R_1, R_2 and T be the graphs shown in Figure 11.1.

   Let $\alpha = (1\ 4\ 2)$ be a permutation of the vertices, and let $\hat{\alpha}$ be the induced permutation on the unordered pairs of vertices. Then $\hat{\alpha}$ is an embedding from R_1 to T. However, there is no embedding of R_2 into T.

8) Let R be the graph $K_2 \cup 2K_1$ and let T be the graph P_4 (the path on four vertices). Then $[T]_R = 12$.

9) Let $R = 2K_2$ and $T = P_4$. Then $[T]_R = 8$.

An embedding of a structure R into itself is called an *automorphism* of R. The set of all automorphisms of R is a subgroup of S_D, and it is denoted by $\mathrm{Aut}(R)$. Note that if R and T are isomorphic structures, then $[R]_T = |\mathrm{Aut}(R)| = |\mathrm{Aut}(T)|$.

The easy proof of the following lemma is left as an exercise.

Lemma 11.1 *Let R and T be two compatible structures. Then*

$$[T]_R = \binom{T}{R}|\mathrm{Aut}(R)|.$$

(This lemma and the previous examples motivate the notation $\binom{T}{R}$ and $[T]_R$ by analogy with the notation for the binomial coefficients and the falling factorial, respectively.)

The following idea of embeddings with forbidden positions will be of central importance in the sequel. Let R, T be two structures and let $X \subseteq \mathcal{E}(R)$. Then $(R \to_X T)$ is the set of all $\alpha \in \Gamma$ such that

(i) if $x \in \mathcal{E}(R) - X$, then $x \in \mathcal{E}(T)$;
(ii) if $x \in X$, then $\alpha(x) \notin \mathcal{E}(T)$.

The size of $(R \to_X T)$ is denoted by $[T]_{R \setminus X}$.

Example

10) Let R be the structure

and let T be the structure

Suppose the cells of the grids are numbered as previously, and let X be the set containing the cells 4 and 7. Then $\alpha = (2\ 4)(3\ 7)(6\ 8)$ is in $(R \to_X T)$.

[*Remarks:* We make a few observations about the very important definition we have just given.

(i) A permutation α in $(R \to_X T)$ maps all of the edges in X into nonedges of T; the other elements of $\mathcal{E}(R)$ are mapped into edges of T. Nonedges of R can be mapped into any type of element of the structure T.

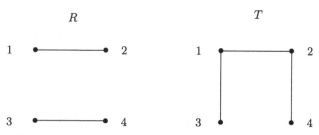

Figure 11.2. Illustrating embeddings with forbidden positions

One can also define the permutations $\alpha \in (R \to_X T)$ as those with the properties

(i) $\alpha(X) \cap \mathcal{E}(T) = \emptyset$;

(ii) $\alpha(\mathcal{E}(R) - X) \subseteq \mathcal{E}(T)$.

(ii) Note that $[T]_{R\backslash X}$ is not equal to $[T]_{R-X}$ because in the latter case we are counting those permutations that map elements of $\mathcal{E}(R - X)$ into elements of $\mathcal{E}(T)$, but in the first case we have the extra restriction that elements of X *must* be mapped into nonedges of T. Note, however, that any element of $(R \to_X T)$ gives a homomorphism from $R - X$ to T, that is, $[T]_{R-X} \geq [T]_{R\backslash X}$.

(iii) It is easy to see that

$$[T]_{R-X} = \sum_{\emptyset \subseteq Y \subseteq X} [T]_{R\backslash Y}.$$

In the next section we shall invert this relation to get $[T]_{R\backslash X}$ as a sum of terms $[T]_{R-Y}$.

(iv) Note that $|R \to_\emptyset T| = [T]_R$.

(v) Suppose $[T]_{R\backslash X} > 0$. Then there is some set $X' \subseteq D - \mathcal{E}(R)$ such that, if the elements of X are made nonedges while those of X' are made edges, then the resulting structure is isomorphic to T, that is,

$$[T]_{R\backslash X} > 0 \Rightarrow \exists X' \in D - \mathcal{E}(R) \text{ such that } R - X + X' \simeq T.]$$

Examples

11) Let R and T be the graphs shown in Figure 11.2, and let X be the set $\{\{3, 4\}\}$ (remember, when considering graphs as structures it is the pairs of vertices that are the elements of D.)

Then $[T]_{R\backslash X} = 4$. Compare this with Example 8.

12) Let R be the structure

and let T be the structure

and let X be the set containing the middle cell in the top row. Then $[T]_{R \setminus X} = 0$, although $[T]_{R-X} > 0$.

11.3 Lovász' and Nash-Williams' Theorems

Theorem 11.2 (Lovász) *Let R, T be two compatible structures and let $X \subseteq \mathcal{E}(R)$. Then*

$$[T]_{R \setminus X} = \sum_{Y \subseteq X} (-1)^{|Y|} [T]_{R-X+Y}.$$

Proof Let $X = \{x_1, \ldots, x_p\}$ and let $S = R - X$. Let A_i denote the set of all embeddings from $S + x_i$ to T. In order to count $[T]_{R \setminus X}$ we have to remove from $[T]_S$ those embeddings that map some element x_i into an element in $\mathcal{E}(T)$. That is,

$$[T]_{R \setminus X} = [T]_S - |A_1 \cup \cdots \cup A_p|.$$

We now apply the inclusion-exclusion principle to $|A_1 \cup \cdots \cup A_p|$. Note that

$$|A_{i_1} \cap A_{i_2} \cap \cdots \cap A_{i_j}| = [T]_{S+Y},$$

where $Y = \{x_{i_1}, x_{i_2}, \ldots, x_{i_j}\}$. Therefore

$$\begin{aligned} [T]_{R \setminus X} &= [T]_S - |A_1 \cup \cdots \cup A_p| \\ &= [T]_S + \sum_{Y \subseteq X, Y \neq \emptyset} (-1)^{|Y|} [T]_{S+Y} \end{aligned}$$

$$= \sum_{Y \subseteq X} (-1)^{|Y|} [T]_{S+Y}$$

$$= \sum_{Y \subseteq X} (-1)^{|Y|} [T]_{R-X+Y}.$$

$\square$

In order to apply this theorem to reconstruction we need the following lemma, which is the analogue, for structures, of Kelly's Lemma for graphs. The proofs of this lemma and its immediate corollary are left as easy exercises.

Lemma 11.3 *Let R, T, S be compatible structures with $\mathcal{SD}(R) = \mathcal{SD}(T)$ and $|\mathcal{E}(S)| < |\mathcal{E}(R)| = |\mathcal{E}(T)|$. Then $\binom{R}{S} = \binom{T}{S}$.*

Corollary 11.4 *Let S, R, T be as in Lemma 11.3 Then $[R]_S = [T]_S$.*

Theorem 11.5 (Nash-Williams) *Let R, T be two compatible structures with $\mathcal{SD}(R) = \mathcal{SD}(T)$, and let $X \subseteq \mathcal{E}(R)$. Then*

$$[T]_R = |\mathrm{Aut}(R)| + (-1)^{|X|} \left([T]_{R \setminus X} - [R]_{R \setminus X} \right).$$

Proof By Lovász' Theorem,

$$[T]_{R \setminus X} = \sum_{Y \subseteq X} (-1)^{|Y|} [T]_{R-X+Y}$$

and

$$[R]_{R \setminus X} = \sum_{Y \subseteq X} (-1)^{|Y|} [R]_{R-X+Y}.$$

Subtracting these two equations, all terms on the right-hand side cancel, by Corollary 11.3, except for $Y = X$, because, when $Y \subset X$, $|\mathcal{E}(R - X + Y)| < |\mathcal{E}(R)|$. The result then follows by noting that $[R]_R = |\mathrm{Aut}(R)|$. $\square$

Corollary 11.6 *Let R, T be two compatible structures with $\mathcal{SD}(R) = \mathcal{SD}(T)$ and suppose $R \not\cong T$. Let $X \subseteq \mathcal{E}(R)$. Then,*

(i) *if $|X|$ is odd, then $[T]_{R \setminus X} > 0$;*
(ii) *if $|X|$ is even, then $[R]_{R \setminus X} > 0$.*

Proof (i) By Nash-Williams' Theorem,

$$[T]_{R \setminus X} - [R]_{R \setminus X} = |\text{Aut}(R)| - [T]_R.$$

But $[T]_R = 0$ because $R \not\simeq T$ and $|\mathcal{E}(R)| = |\mathcal{E}(T)|$. Therefore

$$[T]_{R \setminus X} = |\text{Aut}(R)| + [R]_{R \setminus X} > 0.$$

(ii) Again by Nash-Williams' Theorem,

$$[T]_{R \setminus X} - [R]_{R \setminus X} = [T]_R - |\text{Aut}(R)|.$$

Therefore

$$[R]_{R \setminus X} = |\text{Aut}(R)| + [T]_{R \setminus X} > 0.$$

$\square$

[*Remark.* Case (i) of this corollary therefore says that there is a set X' of nonedges in R such that $R - X + X' \simeq T$, and Case (ii) says that there is such an X' such that $R - X + X' \simeq R$. (See Remark (v) following Example 10.)]

This theory gives the following very important results of Lovász and Müller.

Theorem 11.7 (Lovász) *Let $R = (D, \Gamma, \mathcal{E})$ be a structure and suppose $|\mathcal{E}(R)| > |D|/2$. Then R is $\mathcal{S}$-reconstructible.*

Proof Suppose R is not $\mathcal{S}$-reconstructible. Then there is some reconstruction T of R that is not isomorphic to R. Let $X = \mathcal{E}(R)$. By Case (i) of Corollary 11.6, if X is odd, then there is a set X' of nonedges in R such that $R - X + X' \simeq T$, whereas if $|X|$ is even, then there is such an X' with $R - X + X' \simeq R$. But this is impossible because $|\mathcal{E}(R)| > |D|/2$. $\square$

Theorem 11.8 (Müller) *Let $R = (D, \Gamma, \mathcal{E})$ be a structure and suppose $2^{|\mathcal{E}(R)|-1} > |\Gamma|$. Then R is $\mathcal{S}$-reconstructible.*

Proof Suppose R is not $\mathcal{S}$-reconstructible. By Case (ii) of Corollary 11.6, for every even subset X of $\mathcal{E}(R)$, $[R]_{R \setminus X} \geq 1$. But the sets $(R \to_X R)$ are disjoint for different X. Therefore, since there are $2^{|\mathcal{E}(R)|-1}$ even subsets of $\mathcal{E}(R)$, these sets between them give at least $2^{|\mathcal{E}(R)|-1}$ permutations in $(G \to_X G)$. But this is impossible because these permutations are all in Γ and $2^{|\mathcal{E}(R)|-1} > |\Gamma|$. $\square$

11.4 Extensions

We briefly mention here some ways in which the theory of this chapter can be extended.

The $\mathcal{S}_k$-*deck* of a structure S is the collection of all substructures $S - A$ for all k-subsets A of $\mathcal{E}(S)$. The structure S is said to be $\mathcal{S}_k$-*reconstructible* if it is uniquely (up to isomorphism) determined from its $\mathcal{S}_k$-deck. The following is just one possible extension of the previous results.

Theorem 11.9 (Alon, Caro, Krasikov and Roditty) *Let $R = (D, \Gamma, \mathcal{E})$ be a structure and suppose $2^{|\mathcal{E}(R)|-k} > |\Gamma|$. Then R is $\mathcal{S}_k$-reconstructible.*

The paper of Alon, Caro, Krasikov and Roditty contains several other deeper results of this type and should be studied carefully by anyone who is interested in extending the reconstruction of structures in this direction. Many of the techniques and results these authors developed have been extended more recently by Radcliffe and Scott. They consider the problem of reconstructing a subset of $\mathbb{Z}_n$ (or of $\mathbb{R}$) up to translation from the collection of its subsets of a given size, also given up to translation. The following is a sample of their important results.

Theorem 11.10 (Radcliffe and Scott)

(i) *Suppose n is prime. Then every subset of $\mathbb{Z}_n$ is reconstructible from the collection of its 3-subsets.*

(ii) *For arbitrary n, almost all subsets of $\mathbb{Z}_n$ are reconstructible from the collections of their 3-subsets.*

(iii) *For any n, every subset of $\mathbb{Z}_n$ is reconstructible from its $9\alpha(n)$-subsets, where $\alpha(n)$ is the number of distinct prime factors of n.*

(iv) *A locally finite subset of $\mathbb{R}$ (that is a subset which contains only finitely many translates of any given finite set of size at least 2) is reconstructible from its 3-subsets.*

Looking at edge-reconstruction in this guise, that is, as reconstruction of structures, has led some authors to extend the problem in another direction, namely, by focusing attention more directly on the permutation group involved. Thus, let (Γ, D) be a permutation group. Then the *reconstruction index* $\rho(\Gamma, D)$ of (Γ, D) is the smallest t such that, for any $\mathcal{E} \subset D$ with $|\mathcal{E}| \geq t$, the structure $(D, \Gamma, \mathcal{E})$ is reconstructible. The Edge-Reconstruction Conjecture therefore states that, if $Y = \{1, 2, \ldots, n\}$ and $D = \binom{Y}{2}$, then $\rho(S_n^{(2)}, D) = 4$.

These are some of the results obtained on the reconstruction index of certain groups.

Theorem 11.11 (Mnukhin) *The reconstruction index of all abelian groups is 4 and the reconstruction index of Hamiltonian groups is 5.*

Theorem 11.12 (Maynard) *Let Γ be a semiregular permutation group (that is, a permutation group all of whose elements are semiregular) acting on the set D with $|D| > 6$. Then, $\rho(\Gamma, D)$ is equal to:*

(i) *5 if the quaternion group is a subgroup of Γ;*

(ii) *3 if $|\Gamma|$ is divisible by $2^a \cdot 3 \cdot 5$ for some positive integer a, every Sylow 2-subgroup of Γ is elementary abelian and the centraliser $C_\Gamma(g)$ is equal to $\langle g \rangle$ for every $g \in \Gamma$ of order 3 or 5;*

(iii) *4 otherwise.*

Finally, to consider an extension in another direction (but which ties up with what we did in an earlier chapter on endvertex-deck reconstruction), let us define a structure R to be a triple $R = (D, \Gamma, \mathcal{P})$, where D is a finite set, Γ is a group of permutations of D and $\mathcal{P}$ is a partition $\{\mathcal{E}_0, \mathcal{E}_1, \ldots, \mathcal{E}_n\}$ of D, $n \geq 1$. For any $x \in \mathcal{E}_i$, $i > 0$, the substructure $R - x$ is now defined to be the structure $(D, \Gamma, \mathcal{P}')$, where $\mathcal{P}'$ is now the partition $\{\mathcal{E}_0, \ldots, \mathcal{E}_{i-1} \cup \{x\}, \mathcal{E}_i - x, \ldots, \mathcal{E}_n\}$. Now, R is said to be $\mathcal{S}$-reconstructible if it is uniquely determined from the collection of substructures $R - x$ for all $x \in D - \mathcal{E}_0$.

The proof of the following theorem follows very closely those of Theorems 11.2–11.8.

Theorem 11.13 *Let $R = (D, \Gamma, \mathcal{P})$ be a structure and let t be the size of a largest set in the partition $\mathcal{P}$. If either $t > |D|/2$ or $t > 1 + \lg |\Gamma|$, then R is $\mathcal{S}$-reconstructible.*

This then leads to the following corollary on a problem that we have already discussed, namely, the endvertex-deck reconstruction problem, but where now two or more endvertices can share a common neighbour.

Corollary 11.14 *Let G' be a graph with minimum degree at least 2 and let G be a graph obtained from G' by attaching endvertices to the vertices of G (more than one endvertex may be attached to the same vertex of G'). Supppose there are t vertices of G', each of which is adjacent to the same number of endvertices in G.*

If either $t > |V(G')|/2$ or $t > 1 + \lg |\mathrm{Aut}(G')|$, then G is endvertex-reconstructible.

11.5 Exercises

11.1 Prove Lemma 11.1.

11.2 Prove Lemma 11.3 and Corollary 11.4

11.3 Let $X, Y \subseteq \mathcal{E}(R)$ for some structure R, and let $X \neq Y$. Show that the sets $(R \rightarrow_X T)$ and $(R \rightarrow_Y T)$ are disjoint.

11.4 Let T be the structure consisting of an $n \times n$ grid with 0 and 1 in its cells. Show that, if the number of 1s is more than 4, then T is $\mathcal{S}$-reconstructible. Suppose T is an $n \times n \times n$ cube on whose faces $n \times n$ grids are drawn. Show that if the number of 1s is more than 5, then T is $\mathcal{S}$-reconstructible.

11.5 Let G be a graph and let $n = |V(G)|$, $m = |E(G)|$. Show that, if $m > n(n-1)/4$ or $m > 1 + \lg n!$, then G is edge-reconstructible.

11.6 Consider the nonedge-reconstructible pair $G = P_3 \cup K_1$ and $H = 2K_2$. Verify Corollary 11.6 for this pair. Repeat for the pair of graphs $G = K_3 \cup K_1$ and $H = K_{1,3}$.

11.7 Let the graph G have the property that no six vertices of G induce a 6-cycle. Show that, if G contains K_6 as a subgraph, then G is edge-reconstructible.

11.8 Let G be a graph on at least ten vertices and suppose that it does not contain any induced subgraph isomorphic to the Petersen graph. Show that if G contains K_{10} as a subgraph then G is edge-reconstructible.

11.9 Let H be a graph with minimum degree at least 2 and let G be obtained from H by adding k endvertices and joining them to k distinct vertices of H. Show that, if $k > |V(H)|/2$ or $k > 1 + \lg |\mathrm{Aut}(H)|$, then G is endvertex-reconstructible.

11.10 The following is a slight generalisation of Müller's Theorem, which then follows from $m' = 0$. Let R be a structure and P a substructure of R with $|\mathcal{E}(P)| = m' < m = |\mathcal{E}(R)|$. Show that, if $2^{m-m'-1} > [R]_P$, then R is $\mathcal{S}$-reconstructiible.

11.11 Consider a necklace R made up of N beads equally spaced along a circular wire; m of these beads are coloured black, the others are coloured white. A 'subnecklace' of R is obtained by changing the colour of a black bead into white. Show that, if R has at least six black beads, then it is reconstructible from the deck made up of all of its subnecklaces.

11.12 Let P be a spanning subgraph of G, let $m = |E(G)|$ and $m' = |E(P)|$. Suppose $2^{m-m'-1} > [G]_P$. Prove that G is edge-reconstructible.

11.13 Lovász has shown that a Hamiltonian graph on m edges and n vertices cannot contain more than

$$\frac{n}{2} \left(\frac{m}{n-1} \right)^{n-1}$$

Hamiltonian paths. Let G be a Hamiltonian graph. Using this result, prove that, if

$$2^{m-n} > n \left(\frac{m}{n-1} \right)^{n-1},$$

then G is edge-reconstructible.

11.14 Let T be a spanning tree of the connected graph G having n vertices, and let G have maximum degree Δ. Prove that

$$[G]_T \le n \cdot \Delta! (\Delta - 1)^{n-\Delta-1}.$$

11.15 Show that, if $|V(G)| > 16$, then either G or its complement $\overline{G}$ is reconstructible.

11.16 Call a vertex u of G *irreplaceable* if for *no* set of neighbours A of u is there a set B of vertices of G, not adjacent to u, such that G is isomorphic to the graph obtained by removing all edges in $\{ua : a \in A\}$ and replacing them by edges $\{ub : b \in B\}$.

Show that an irreplaceable vertex u of G cannot be pseudosimilar to any vertex to which it is not adjacent. Also show that, if $\deg(u) \geq 2$ and u has minimum degree in G, then G is edge-reconstructible.

11.17 Look up the proofs of Theorems 11.9 and 11.13 and Corollary 11.14.

11.6 Notes and guide to references

The methods discussed in this chapter originated from Lovász' paper [151] and Müller's sequel [196]. The formulation of Nash-Williams' Lemma appeared in [203]. Perhaps the most striking use of the methods in this chapter in obtaining results in the edge-reconstruction of graphs has been accomplished by Pyber [217], who has shown that any Hamiltonian graph with a sufficiently large number of vertices is edge-reconstructible. The upper bound on the number of Hamiltonian paths in a Hamiltonian graph was obtained in [152].

A recent paper that collects results of the type obtained in this chapter, that is, involving the automorphism group of a graph and Nash-Williams' Lemma, is [133].

Extensions of these methods to structures other than graphs in the manner treated in this chapter, including the study of the reconstruction index of a permutation group, have appeared in [3, 46, 47, 178, 179, 180, 188, 189, 190, 191, 218, 219].

Theorem 11.9 is just a little sample of the many results obtained in this direction in [3]; and [218, 219], which contain Theorem 11.10, give very important extensions of a number of the results in [3]. The reconstruction index of a permutation group is becoming a topic of considerable interest, and new results in this field should be expected. Theorems 11.11 and 11.12 can be found in [191] and [179], respectively. The proofs of all of these results are long and sometimes very deep, and so could not be reproduced here. However, one scope of this chapter, and the reason edge-reconstruction has been presented as the reconstruction of 'structures' with a group acting on them, rather than simply the edge-reconstruction of graphs, was to bring the reader to the point where he can begin to understand these works and possibly participate in the search for new results that they invite.

Theorem 11.13 is a relatively simple extension of the ideas contained in this chapter, but its corollary ties up with earlier results on endvertex-deck reconstruction. These proofs can be found in [137].

References

[1] U. P. Acharya and H. S. Mehta. "2-Cartesian product of special graphs". In: *Int. J. Math. and Soft Comput.* 4.1 (2014), pp. 139–144 (cit. on p. 108).

[2] Á. Ádám. "Research problem 2-10". In: *J. Combin. Theory* 2 (1967), p. 393 (cit. on p. 122).

[3] N. Alon, Y. Caro, I. Krasikov, and Y. Roditty. "Combinatorial reconstruction problems". In: *J. Combin. Theory (Ser. B)* 47 (1989), pp. 153–161 (cit. on p. 169).

[4] N. Alon and J. H. Spencer. *The Probabilistic Method.* Wiley, 1992 (cit. on p. 38).

[5] B. Alspach. "Point-symmetric graphs and digraphs of prime order and transitive groups of prime degree". In: *J. Combin. Theory (Ser. B)* 15 (1973), pp. 12–17 (cit. on p. 122).

[6] B. Alspach. "Isomorphism and Cayley graphs on abelian groups". In: *Graph Symmetry: Algebraic Methods and Applications.* Ed. by G. Hahn and G. Sabidussi. Kluwer Acad. Publ., 1997, pp. 1–22 (cit. on p. 122).

[7] B. Alspach, D. Marušič, and L. A. Nowitz. "Constructing graphs which are 1/2-transitive". In: *J. Austral. Math. Soc. (Ser. A)* 56 (1994), pp. 391–402 (cit. on p. 48).

[8] B. Alspach and T. D. Parsons. "A construction for vertex-transitive graphs". In: *Canad. J. Math.* 34 (1982), pp. 307–318 (cit. on pp. 115, 122).

[9] B. Alspach and M.-Y. Xu. "$\frac{1}{2}$-arc-transitive graphs of order $3p$". In: *J. Algebraic Combin.* 3 (1994), pp. 347–355 (cit. on p. 91).

[10] W. C. Arlinghaus and F. Harary. "The digraph number of a finite abelian group". In: *Wiss. Z. Tech. Hochsch. Ilmenau* 33.1 (1987), pp. 25–31 (cit. on p. 62).

[11] M. Aschbacher. "The nonexistence of rank three permutation groups of degree 3250 and subdegree 57". In: *J. Algebra* 19 (1971), pp. 538–540 (cit. on p. 77).

[12] K. Asciak. "The degree-associated edge-reconstruction number of disconnected graphs and trees". Preprint. 2015 (cit. on p. 144).

[13] K. Asciak and J. Lauri. "On disconnected graphs with large reconstruction numbers". In: *Ars Combin.* 62 (2002), pp. 173–181 (cit. on p. 143).

[14] K. Asciak and J. Lauri. "On the edge-reconstruction number of disconnected graphs". In: *Bull. Inst. Combin. and Its Applics.* 63 (2011), pp. 87–100 (cit. on p. 143).

171

[15]   K. Asciak, J. Lauri, W. Myrvold, and V. Pannone. "On the edge-reconstruction number of a tree". In: *Australas. J. Combin.* 60 (2014), pp. 169–190 (cit. on p. 144).

[16]   L. Babai. "Long cycles in vertex-transitive graphs". In: *J. Graph Theory* 3 (1979), pp. 23–29 (cit. on p. 60).

[17]   L. Babai. "On the abstract group of automorphisms". In: *Combinatorics.* Ed. by H. N. V. Temperley. Vol. 52. London Math. Soc. Lecture Note Series. Proceedings of the Eighth British Combinatorial Conference University College Swansea, 1981. Cambridge University Press, 1981, pp. 1–40 (cit. on p. 62).

[18]   L. Babai. "Automrphism groups, isomorphism, reconstruction". In: *Handbook of Combinatorics.* Ed. by R. Graham, M. Grötschel, and L. Lovász. Vol. 2. Elsevier Science B.V., 1995. Chap. 27, pp. 1447–1540 (cit. on p. 63).

[19]   D. W. Bange, A. E. Barkauskas, and L. H. Host. "Class-reconstruction of total graphs". In: *J. Graph Theory* 11 (1987), pp. 221–230 (cit. on p. 147).

[20]   M. D. Barrus and D. B. West. "Degree associated reconstruction number of graphs". In: *Discrete Math.* 310 (2010), pp. 2600–2612 (cit. on p. 144).

[21]   R. A. Beaumont and R. P. Peterson. "Set-transitive permutation groups". In: *Canadian J. Math.* 7 (1955), pp. 35–42 (cit. on p. 115).

[22]   M. Behzad, G. Chartrand, and L. Lesniak-Foster. *Graphs and Digraphs.* Prindle, Weber & Schmidt, 1979 (cit. on pp. 17, 60, 62).

[23]   L. W. Beineke and E. T. Parker. "On nonreconstructable tournaments". In: *J. Combinatorial Theory* 9 (1970), pp. 324–326 (cit. on p. 137).

[24]   N. L. Biggs. *Algebraic Graph Theory.* Cambridge University Press, 1993 (cit. on pp. 17, 32, 34, 38, 77, 108, 154).

[25]   N. L. Biggs and D. H. Smith. "On trivalent graphs". In: *Bull. London Math. Soc.* 3 (1971), pp. 155–158 (cit. on p. 38).

[26]   N. L. Biggs and A. T. White. *Permutation Groups and Combinatorial Structures.* Cambridge University Press, 1979 (cit. on p. 77).

[27]   B. Bollobás. "Almost every graph has reconstruction number 3". In: *J. Graph Theory* 14 (1990), pp. 1–4 (cit. on p. 38).

[28]   B. Bollobás. *Modern Graph Theory.* Springer-Verlag, 1998 (cit. on p. 17).

[29]   B. Bollobás. *Random Graphs.* Cambridge University Press, 2001 (cit. on pp. 36, 38).

[30]   A. Bondy and U. S. R. Murty. *Graph Theory (Graduate Texts in Mathematics).* Springer, 2008 (cit. on p. 36).

[31]   J. A. Bondy. "A graph reconstructor's manual". In: *Surveys in Combinatorics.* Ed. by A. D. Keedwell. Cambridge University Press, 1991, pp. 221–252 (cit. on pp. 123, 156).

[32]   J. A. Bondy and R. L. Hemminger. "Graph reconstruction—a survey". In: *J. Graph Theory* 1 (1977), pp. 227–268 (cit. on pp. 123, 137, 139).

[33]   I. Z. Bouwer. "An edge but not vertex transitive cubic graph". In: *Canad. Math. Bull.* 11.4 (1968), pp. 533–534 (cit. on p. 20).

[34]   I. Z. Bouwer. "Vertex and edge transitive but not 1-transitive graphs". In: *Canad. Math. Bull.* 13 (1970), pp. 231–237 (cit. on p. 48).

[35]   I. Z. Bouwer. "Section graphs for finite permutation groups". In: *J. Combin. Theory* 6 (1971), pp. 378–386 (cit. on p. 62).

[36] D. P. Bovet and P. Crescenzi. *Introduction to the Theory of Complexity*. Prentice Hall, 1994 (cit. on pp. 13, 17).

[37] A. Bowler, P. A. Brown, and T. Fenner. "Families of pairs of graphs with a large number of common cards". In: *J. Graph Theory* 63.2 (2010), pp. 146–163 (cit. on p. 144).

[38] A. Bowler, P. A. Brown, T. Fenner, and W. Myrvold. "Recognizing connectedness from vertex-deleted subgraphs". In: *J. Graph Theory* 67.4 (2011), pp. 285–299 (cit. on p. 144).

[39] A. E. Brouwer. *Parameters of Strongly Regular Graphs*. www.win.tue.nl/~aeb/graphs/srg/srgtab.html (cit. on pp. 76, 77).

[40] A. E. Brouwer, A. M. Cohen, and A. Neumaier. *Distance-Regular Graphs*. Springer-Verlag, 1989 (cit. on p. 38).

[41] P. A. Brown. "On the Maximum Number of Common Cards between Various Classes of Graphs". PhD thesis. Birkbeck College, University of London, 2008 (cit. on p. 144).

[42] J. M. Burns and B. Goldsmith. "The trace of an abelian group—an application to digraphs". In: *Proc. Roy. Irish. Acad. Sect. A* 95 (1995), pp. 75–79 (cit. on p. 62).

[43] P. J. Cameron. "Strongly regular graphs". In: *Selected Topics in Graph Theory*. Ed. by L. W. Beineke and R. J. Wilson. Academic Press, 1978. Chap. 12 (cit. on p. 77).

[44] P. J. Cameron. "Automorphism groups of graphs". In: *Selected Topics in Graph Theory, Vol. 2*. Ed. by L. W. Beineke and R. J. Wilson. Academic Press, 1983. Chap. 4 (cit. on p. 77).

[45] P. J. Cameron. *Oligomorphic Permutation Groups*. Cambridge University Press, 1990 (cit. on p. 63).

[46] P. J. Cameron. "Some open problems on permutation groups". In: *Groups, Combinatorics and Geometry*. Ed. by M. W. Liebeck and J. Saxl. London Mathematical Society Lecture Notes 165. Cambridge University Press, 1992 (cit. on p. 169).

[47] P. J. Cameron. "Stories from the age of reconstruction". In: *Congr. Num.* 113 (1996), pp. 31–41 (cit. on p. 169).

[48] P. J. Cameron. "Oligomorphic groups and homogeneous graphs". In: *Graph Symmetry: Algebraic Methods and Its Applications*. Ed. by G. Hahn and G. Sabidussi. Kluwer Acad. Publ., 1997, pp. 23–74 (cit. on p. 63).

[49] P. J. Cameron. *Permutation Groups*. Vol. 45. London Mathematical Society Student Texts. Cambridge University Press, 1999 (cit. on pp. 7, 17, 38, 63, 77).

[50] P. J. Cameron and J. H. van Lint. *Designs, Graphs, Codes and Their Links*. Vol. 22. London Mathematical Society Student Texts. Cambridge University Press, 1991 (cit. on p. 77).

[51] M. Capobianco and J. C. Molluzzo. *Examples and Counterexamples in Graph Theory*. North-Holland, 1978 (cit. on p. 122).

[52] K. M. Cattermole. "Graph theory and connection networks". In: *Applications of Graph Theory*. Ed. by R. J. Wilson and L. W. Beineke. Academic Press, 1979, pp. 17–57 (cit. on p. 97).

[53] P. V. Ceccherini and A. Sappa. "A new characterization of hypercubes". In: *Ann. Discrete Math.* 30 (1986), pp. 137–142 (cit. on p. 38).

[54] G. Chartrand, A. Kaugars, and D. R. Lick. "Critically n-connected graphs". In: *Proc. Amer. Math. Soc.* 32 (1972), pp. 63–68 (cit. on p. 139).

[55] P. Z. Chinn. "A graph with p points and enough distinct $(p − 2)$-order subgraphs is reconstructible". In: *Recent Trends in Graph Theory*. Ed. by M. Capobianco et al. Vol. 186. Lecture Notes in Mathematics. Springer-Verlag, 1971, pp. 71–73 (cit. on p. 139).

[56] M. Conder, A. Malnič, D. Marušič, T. Pisanski, and P. Potočnik. "The edge-transitive but not vertex-transitive cubic graph on 112 vertices". In: *J. Graph Theory* 50.1 (2005), pp. 25–42 (cit. on p. 38).

[57] H. S. M. Coxeter, R. Frucht, and D. L. Powers. *Zero-Symmetric Graphs: Trivalent Graphical Regular Representations of Groups*. Academic Press, 1981 (cit. on pp. 91, 122).

[58] D. M. Cvetković, M. Doob, and H. Sachs. *Spectra of Graphs (3rd Ed.)* Johann Ambrosius Barth, 1995 (cit. on p. 38).

[59] D. Cvetković and M. Lepović. "Seeking counterexamples to the reconstruction conjecture for the characteristic polynomial of graphs and a positive result". In: *Bull. Cl. Sci. Math. Nat. Sci. Math.* 23 (1998), pp. 91–100 (cit. on p. 148).

[60] E. R. van Dam. "Nonregular graphs with three eigenvalues". In: *J. Combin. Theory Ser. B* 73.2 (1998), pp. 101–118 (cit. on p. 73).

[61] R. Diestel. *Graph Theory*. Springer-Verlag, 1997 (cit. on p. 17).

[62] J. D. Dixon and B. Mortimer. *Permutation Groups*. Springer-Verlag, 1996 (cit. on pp. 7, 17).

[63] W. Dörfler. "Every regular graph is a quasigroup graph". In: *Discrete Math.* 10 (1974), pp. 181–183 (cit. on p. 122).

[64] P. Doyle. "A 27-vertex graph that is vertex-transitive and edge-transitive but not 1-transitive". In: URL: http://arxiv.org/abs/math/0703861(1998) (cit. on p. 48).

[65] P. Dulio and V. Pannone. "The converse of Kelly's Lemma and control-classes in graph reconstruction". In: *Acta Univ. Palack. Olomuc. Fac. Rerum Natur. Math.* 44 (2005), pp. 25–38 (cit. on pp. 129, 139).

[66] B. Elspas and J. Turner. "Graphs with circulant adjacency matrices". In: *J. Combin. Theory* 9 (1970), pp. 297–307 (cit. on p. 122).

[67] P. Erdös, C. Ko, and R. Rado. "Intersection theorems for systems of finite sets". In: *Quart. J. Math.* 12 (1961), pp. 313–320 (cit. on p. 122).

[68] P. Erdös, A. Rényi, and V. T. Sós. "On a problem of graph theory". In: *Studia Sci. Math. Hungar.* 1 (1966), pp. 215–235 (cit. on p. 77).

[69] G. Exoo. *Miscellaneous Topics in Combinatorics*. URL: http://ginger.indstate.edu/ge/COMBIN/index.html (cit. on pp. 78, 87).

[70] H. Fan. "Edge reconstruction of planar graphs with minimum degree at least three—IV". In: *Systems Sci. Math. Sci.* 7 (1994), pp. 218–222 (cit. on p. 139).

[71] Xin Gui Fang, Cai Heng Li, Jie Wang, and Ming Yao Xu. "On cubic Cayley graphs of finite simple groups". In: *Discrete Math.* 244.1-3 (2002). Algebraic and topological methods in graph theory (Lake Bled, 1999), pp. 67–75 (cit. on p. 91).

[72] I. A. Faradzev and M. Klin. "Computer package for computation with coherent configurations". In: Proc. ISSAC-91. (Bonn). ACM Press, 1991, pp. 219–223 (cit. on p. 77).

[73]   S. Fiorini. "A theorem on planar graphs with an application to the reconstruction problem, I". In: *Quart. J. Math. Oxford (2)* 29 (1978), pp. 353–361 (cit. on p. 134).

[74]   S. Fiorini. "On the edge-reconstruction of planar graphs". In: *Math. Proc. Camb. Phil. Soc.* 83 (1978) (cit. on p. 139).

[75]   S. Fiorini and J. Lauri. "The reconstruction of maximal planar graphs. I. Recognition". In: *J. Combin. Theory Ser. B* 30.2 (1981), pp. 188–195 (cit. on p. 135).

[76]   S. Fiorini and B. Manvel. "A theorem on planar graphs with an application to the reconstruction problem. II". In: *J. Combin. Inform. System Sci.* 3.4 (1978), pp. 200–216 (cit. on p. 135).

[77]   J. Folkman. "Regular line-symmetric graphs". In: *J. Combin. Theory* 3 (1967), pp. 215–232 (cit. on pp. 20, 35).

[78]   R. Frucht. "Graphs of degree three with a given abstract group". In: *Canad. J. Math.* 1 (1949), pp. 365–378 (cit. on p. 62).

[79]   R. Frucht. "How to describe a graph". In: *Ann. N. Y. Acad. Sci.* 175 (1970), pp. 159–67 (cit. on p. 122).

[80]   R. Frucht, J. E. Graver, and M. E. Watkins. "The groups of generalized Petersen graphs". In: *Proc. Cambridge Phil. Soc.* 70 (1971), pp. 211–218 (cit. on pp. 110, 122).

[81]   G. Gamble and C. E. Praeger. "Vertex-primitive groups and graphs of order twice the product of two distinct odd primes". In: *J. Group Theory* 3.3 (2000), pp. 247–269 (cit. on p. 62).

[82]   M. R. Garey and D. S. Johnson. *Computers and Intractability: A Guide to the Theory of NP-Completeness.* W. H. Freeman, 1979 (cit. on pp. 13, 17).

[83]   G. Gauyacq. "Routages uniformes dans les graphes sommet-transitifs". Thèse. Univ. Bordeaux I, 1995 (cit. on p. 117).

[84]   G. Gauyacq. "On quasi-Cayley graphs". In: *Discrete Appl. Math.* 77 (1997), pp. 43–58 (cit. on p. 117).

[85]   C. D. Godsil. "More odd graph theory". In: *Discrete Math.* 32 (1980), pp. 205–207 (cit. on p. 122).

[86]   C. D. Godsil. "Neighborhoods of transitive graphs and GRRs". In: *J. Combin. Theory (Ser. B)* 29 (1980), pp. 116–140 (cit. on p. 91).

[87]   C. D. Godsil. "The automorphism groups of some cubic Cayley graphs". In: *European J. Combin.* 4.1 (1983), pp. 25–32 (cit. on p. 91).

[88]   C. D. Godsil and W. L. Kocay. "Constructing graphs with pairs of pseudo-similar vertices". In: *J. Combin. Theory (Ser. B)* 32 (1982), pp. 146–155 (cit. on p. 91).

[89]   C. D. Godsil and B. D. McKay. "Spectral conditions for the reconstructibility of a graph". In: *J. Combin. Theory (Ser. B)* 30 (1981), pp. 285–289 (cit. on p. 148).

[90]   C. Godsil and G. Royle. *Algebraic Graph Theory.* Springer-Verlag, 2001 (cit. on p. 17).

[91]   D. L. Greenwell. "Reconstructing graphs". In: *Proc. Amer. Math. Soc.* 30 (1971), pp. 431–433 (cit. on p. 139).

[92]   D. L. Greenwell and R. L. Hemminger. "Reconstructing graphs". In: *The Many Facets of Graph Theory.* Ed. by G. Chartrand and S. F. Kapoor. Vol. 110. Lecture Notes in Mathematics. (Proc. of the conference held at Western Michigan

University Kalamazoo Mich., 1968). Springer-Verlag, 1969, pp. 91–114 (cit. on p. 139).

[93] M. Gromov. "Groups of polynomial growth and expanding maps". In: *Inst. Hautes Études Sci. Publ. Math.* 53 (1981), pp. 53–73 (cit. on p. 63).

[94] G. Hahn and G. Sabidussi (Eds.) *Graph Symmetry: Algebraic Methods and Applications.* Kluwer Acad. Publ., 1997 (cit. on p. 17).

[95] G. Hahn and C. Tardif. "Homomorphisms of graphs". In: *Graph Symmetry: Algebraic Methods and Applications.* Ed. by G. Hahn and G. Sabidussi. Kluwer Acad. Publ., 1997, pp. 107–166 (cit. on p. 108).

[96] R. Hammack, W. Imrich, and S. Klavžar. *Handbook of Product Graphs, Second Edition.* Discrete Mathematics and Its Applications. Taylor & Francis, 2011 (cit. on p. 108).

[97] F. Harary. *Graph Theory.* Addison-Wesley, 1969 (cit. on p. 17).

[98] F. Harary and J. Lauri. "The class-reconstruction number of maximal planar graphs". In: *Graphs and Combinatorics* 3 (1987), pp. 45–53 (cit. on pp. 144, 147).

[99] F. Harary and J. Lauri. "On the class-reconstruction number of trees". In: *Quart. J. Math. Oxford (2)* 39 (1988), pp. 47–60 (cit. on pp. 144, 147).

[100] F. Harary and E. M. Palmer. "A note on similar points and similar lines in a graph". In: *Rev. Roum. Math. Pures et Appl* 10 (1965), pp. 1489–1492 (cit. on p. 91).

[101] F. Harary and E. M. Palmer. "On similar points of a graph". In: *J. Math. Mech.* 15 (1966), pp. 623–630 (cit. on p. 91).

[102] F. Harary and E. M. Palmer. "On the problem of reconstructing a tournament from subtournaments". In: *Monatsh. Math.* 71 (1967), pp. 14–23 (cit. on p. 136).

[103] F. Harary and E. M. Palmer. *Graphical Enumeration.* Academic Press, 1973 (cit. on pp. 10, 36, 38).

[104] F. Harary, A. Vince, and D. Worley. "A point-symmetric graph that is nowhere reversible". In: *Siam J. Alg. Disc. Meth.* 3.3 (1982), pp. 285–287 (cit. on p. 91).

[105] D. Hetzel. "Über reguläre graphische Darstellungen von auflösbaren Gruppen". Diplomarbeit. Technische Universität Berlin, 1976 (cit. on p. 91).

[106] C. M. Hoffman. "Subcomplete generalisations of graph isomorphism". In: *J. Computer and System Sciences* 25 (1982), pp. 332–359 (cit. on pp. 17, 63).

[107] D. F. Holt. "A graph which is edge transitive but not arc transitive". In: *J. Graph Theory* 5 (1981), pp. 201–204 (cit. on pp. 47, 48).

[108] D. A. Holton and J. Sheehan. *The Petersen Graph.* Vol. 7. Australian Mathematical Society Lecture Series. Cambridge University Press, 1993 (cit. on pp. 62, 122).

[109] W. Imrich. "Graphs with transitive abelian automorphism group". In: *Combinatorial Theory and Its Applications II.* Ed. by P. Erdős, A. Rényi, and V. T. Sós. Vol. 4. Colloq. Math. Soc. J. Bolyai. North-Holland, 1970, pp. 651–656 (cit. on pp. 89, 91).

[110] W. Imrich. "Assoziative Produkte von Graphen". In: *Osterreich Akad. Wiss. Math.-Natur. Kl. S.-B.* 180.II (1972), pp. 203–239 (cit. on p. 108).

[111] W. Imrich. "On graphs and regular groups". In: *J. Combin. Theory (Ser. B)* 19 (1975), pp. 174–180 (cit. on p. 91).

[112] W. Imrich. "Graphical regular representations of groups of odd order". In: *Combinatorics*. Ed. by A. Hajnal and V. T. Sós. Vol. 18. Colloq. Math. Soc. J. Bolyai. North-Holland, 1976, pp. 611–622 (cit. on p. 91).

[113] W. Imrich and H. Izbichi. "Associative products of graphs". In: *Monatsh. Math.* 80.4 (1975), pp. 277–281 (cit. on pp. 97, 108).

[114] W. Imrich and S. Klavžar. *Product Graphs: Structure and Recognition*. Wiley, 2000 (cit. on p. 108).

[115] W. Imrich and M. E. Watkins. "On graphical regular representations of cyclic extensions of groups". In: *Pacific J. Math.* 55.2 (1974), pp. 461–477 (cit. on p. 91).

[116] W. Imrich and M. E. Watkins. "On automorphism groups of Cayley graphs". In: *Periodica Mathematica Hungarica* 7.3–4 (1976), pp. 243–258 (cit. on p. 91).

[117] T. R. Jensen and B. Toft. *Graph Coloring Problems*. J. Wiley and Sons, 1995 (cit. on p. 105).

[118] I. N. Kagno. "Linear graphs of degree ≤ 6 and their groups". In: *Amer. J. Math.* 68 (1946), pp. 505–520 (cit. on p. 59).

[119] W. Kantor. "*k*-Homogeneous graphs". In: *Math. Z.* 124 (1972), pp. 261–265 (cit. on p. 114).

[120] P. J. Kelly. "A congruence theorem for trees". In: *Pacific J. Math.* 7 (1957), pp. 961–968 (cit. on p. 140).

[121] R. J. Kimble, A. J. Schwenk, and P. K. Stockmeyer. "Pseudosimilar vertices in a graph". In: *J. Graph Theory* 5 (1981), pp. 171–181 (cit. on p. 91).

[122] D. G. Kirkpatrick, M. M. Klawe, and D. G. Corneil. "On pseudosimilarity in trees". In: *J. Combin. Theory (Ser. B)* 34 (1983), pp. 323–339 (cit. on p. 91).

[123] M. H. Klin and R. Pöschel. "The König problem, the isomorphism problem for cyclic graphs and the method of Schur rings". In: *Algebraic Methods in Graph Theory*. Ed. by L. Lovász and V. T. Sós. Vol. 25. Colloq. Math. Soc. J. Bolyai. North-Holland, 1981, pp. 405–430 (cit. on p. 122).

[124] M. Klin, J. Lauri, and M. Ziv-Av. "Links between two semisymmetric graphs on 112 vertices via association schemes". In: *J. Symbolic Comput.* 47.10 (2012), pp. 1175–1191 (cit. on p. 38).

[125] M. Klin, C. Pech, S. Reichard, A. Woldar, and M. Ziv-Av. "Examples of computer experimentation in algebraic combinatorics". In: *Ars Math. Contemp.* 3 (2010), pp. 237–258 (cit. on pp. 48, 77).

[126] J. Köbler, U. Schöning, and J. Torán. *The Graph Isomorphism Problem: Its Structural Complexity*. Birkhäuser, 1993 (cit. on pp. 14, 17, 139).

[127] W. L. Kocay. "On reconstructing spanning subgraphs". In: *Ars Combinatoria* 11 (1981), pp. 301–313 (cit. on p. 156).

[128] W. L. Kocay. "Some new methods in reconstruction theory". In: *Combinatorial Mathematics IX*. Ed. by E. J. Billington, S. Oates-Williams, and A. Penfold Street. Vol. 952. Lecture Notes in Mathematics. (Proc. 9th Australian Conf. on Combinatorial Mathematics, Univ. of Queensland, Brisbane). Springer-Verlag, 1982, pp. 89–114 (cit. on p. 156).

[129] W. L. Kocay. "Attaching graphs to pseudosimilar vertices". In: *J. Austral. Math. Soc. (Ser. A)* 36 (1984), pp. 53–58 (cit. on p. 91).

[130] W. L. Kocay. "On Stockmeyer's non-reconstructible tournaments". In: *J. Graph Theory* 9 (1985), pp. 473–476 (cit. on p. 137).

[131]   W. L. Kocay. "Graphs & groups, a Macintosh application for graph theory". In: *J. Combin. Maths. and Combin. Comput.* 3 (1988), pp. 195–206 (cit. on pp. 14, 17, 63, 139).

[132]   A. D. Korshunov. "Number of nonisomorphic graphs in an *n*-point graph". In: *Math. Notes of the Acad. USSR* 9 (1971), pp. 155–160 (cit. on p. 38).

[133]   I. Krasikov, A. Lev, and B. D. Thatte. "Upper bounds on the automorphism group of a graph". In: *Discrete Math.* 256.1-2 (2002), pp. 489–493 (cit. on p. 169).

[134]   D. Kratsch and L. A. Henaspaandra. "On the complexity of graph reconstruction". In: *Math. Systems Theory* 27.3 (1994), pp. 257–273 (cit. on p. 139).

[135]   V. Krishnamoorthy and K. R. Parthasarathy. "Cospectral graphs and digraphs with given automorphism group". In: *J. Combin. Theory (Ser. B)* 19 (1975), pp. 204–213 (cit. on p. 91).

[136]   J. Lauri. "The reconstruction of maximal planar graphs, II: Reconstruction". In: *J. Combin. Theory (Ser. B.)* 30 (1981), pp. 196–214 (cit. on pp. 136, 139).

[137]   J. Lauri. "Endvertex-deleted subgraphs". In: *Ars Combinatoria* 36 (1993), pp. 171–182 (cit. on pp. 91, 169).

[138]   J. Lauri. "Pseudosimilarity in graphs—A survey". In: *Ars Combinatoria* 36 (1997), pp. 171–182 (cit. on p. 91).

[139]   J. Lauri. "Constructing graphs with several pseudosimilar vertices or edges". In: *Discrete Math.* 267.1-3 (2003). Combinatorics 2000 (Gaeta), pp. 197–211 (cit. on p. 91).

[140]   J. Lauri. "The Reconstruction Problem". In: *Handbook of Graph Theory*. Ed. by J. L. Gross, J. Yellen, and P. Zhang. Discrete Mathematics and Its Applications. 2014 (cit. on p. 123).

[141]   J. Lauri, R. Mizzi, and R. Scapellato. "Two-fold orbital digraphs and other constructions." In: *International J. of Pure and Applied Math.* 1 (2004), pp. 63–93 (cit. on p. 108).

[142]   J. Lauri, R. Mizzi, and R. Scapellato. "Two-fold automorphisms of graphs." In: *Australasian J. Combinatorics.* 49 (2011), pp. 165–176 (cit. on p. 108).

[143]   J. Lauri, R. Mizzi, and R. Scapellato. "A generalisation of isomorphisms with applications." Preprint. 2014 (cit. on pp. 103, 104, 108).

[144]   J. Lauri, R. Mizzi, and R. Scapellato. "A smallest unstable asymmetric graph and an infinite family of asymmetric graphs with arbitrarily large instability index". Preprint. 2015 (cit. on p. 108).

[145]   J. Lauri, R. Mizzi, and R. Scapellato. "Unstable graphs: A fresh outlook via TF-automorphisms". In: *Ars Mathematica Contemporanea* 8.1 (2015), pp. 115–131 (cit. on p. 108).

[146]   J. Lauri and R. Scapellato. "A note on graphs all of whose edges are pseudosimilar". In: *Graph Theory Notes of New York* 21 (1996), pp. 11–13 (cit. on p. 91).

[147]   W. Lederman and A. J. Weir. *Introduction to Group Theory (2nd Ed.)* Longman, 1996 (cit. on pp. 7, 17).

[148]   Cai Heng Li. "The solution of a problem of Godsil on cubic Cayley graphs". In: *J. Combin. Theory Ser. B* 72.1 (1998), pp. 140–142 (cit. on p. 91).

[149]   D. Livingstone and A. Wagner. "Transitivity of finite permutation groups". In: *Math. Z.* 90 (1965), pp. 393–403 (cit. on pp. 114, 115).

[150]   L. Lovász. "Unsolved problem II". In: *Combinatorial Structures and Their Applications.* Ed. by R. Guy, H. Hanani, N. Sauer, and J. Schonheim. Proceedings of the Calgary International Conference on Combinatorial Structures and Their Applications, 1969. Gordan and Breach, 1970 (cit. on p. 62).

[151]   L. Lovász. "A note on the line reconstruction problem". In: *J. Combin. Theory (Ser. B)* 13 (1972), pp. 309–310 (cit. on p. 169).

[152]   L. Lovász. "Some problems of graph theory". In: *Matematikus Kurir* (1983) (cit. on p. 169).

[153]   L. Lovász. *Combinatorial Problems and Exercises.* Second ed. North-Holland Publishing Co., Amsterdam, 1993 (cit. on p. 138).

[154]   M. Lovrečič-Saražin. "A note on the generalized Petersen graphs that are also Cayley graphs". In: *J. Comb. Theory (Ser. B)* 69 (1997), pp. 189–192 (cit. on pp. 111, 122).

[155]   E. M. Luks. "Isomorphism of graphs of bounded valence can be tested in polynomial time". In: *J. Computer and System Sciences* 25 (1982), pp. 42–65 (cit. on p. 17).

[156]   M. Mačaj and J. Širáň. "Search for properties of the missing Moore graph". In: *Linear Alg. and Applics.* 432 (2010), pp. 2381–2389 (cit. on p. 77).

[157]   D. Macpherson. "The action of an infinite permutation group on the unordered subsets of a set". In: *Proc. London Math. Soc.* 46.3 (1983), pp. 471–486 (cit. on p. 63).

[158]   D. Macpherson. "Growth rates in infinite graphs and permutation groups". In: *Proc. London Math. Soc.* 51.3 (1985), pp. 285–294 (cit. on p. 63).

[159]   W. Magnus, A. Karrass, and D. Solitar. *Combinatorial group theory.* Second ed. Presentations of groups in terms of generators and relations. Dover Publications, 2004 (cit. on p. 7).

[160]   A. Malnič, D. Marušič, P. Potočnik, and C. Wang. "An infinite family of cubic edge- but not vertex-transitive graphs". In: *Discrete Math.* 280.1-3 (2004), pp. 133–148 (cit. on p. 20).

[161]   B. Manvel. "Reconstruction of trees". In: *Canadian J. Math.* 22 (1970), pp. 55–60 (cit. on p. 148).

[162]   B. Manvel. "On reconstructing graphs from their sets of subgraphs". In: *J. Combin. Theory (Ser. B)* 21 (1976), pp. 156–165 (cit. on pp. 139, 148).

[163]   D. Marušič. "Cayley properties of vertex symmetric graphs". In: *Ars Combin.* 16B (1983), pp. 297–302 (cit. on p. 62).

[164]   D. Marušič. "Hamiltonian circuits in Cayley graphs". In: *Discrete Math.* 46 (1983), pp. 49–54 (cit. on p. 62).

[165]   D. Marušič. "On vertex-transitive graphs of order qp". In: *J. Combin. Math. Combin. Comput.* 4 (1988), pp. 97–114 (cit. on p. 116).

[166]   D. Marušič and T. Pisanski. "The Gray graph revisited". In: *J. Graph Theory* 35.1 (2000), pp. 1–7 (cit. on p. 21).

[167]   D. Marušič and R. Scapellato. "A class of non-Cayley vertex-transitive graphs associated with $PSL(2, p)$". In: *Discrete Math.* 109 (1992), pp. 161–170 (cit. on p. 122).

[168]   D. Marušič and R. Scapellato. "Characterizing vertex-transitive pq-graphs with an imprimitive automorphism subgroup". In: *J. Graph Theory* 16 (1992), pp. 375–387 (cit. on p. 122).

[169]   D. Marušič and R. Scapellato. "Imprimitive representations of $SL(2, 2^k)$". In: *J. Combin. Theory (Ser. B)* 58 (1993), pp. 46–57 (cit. on pp. 116, 122).

[170]   D. Marušič and R. Scapellato. "A class of graphs arising from the action of $PSL(2, q^2)$ on cosets of $PGL(2, q)$". In: *Discrete Math.* 134 (1994), pp. 99–110 (cit. on p. 122).

[171]   D. Marušič and R. Scapellato. "Classification of vertex-transitive pq-digraphs". In: *Atti Ist. Lombardo (Rend. Sci.)* A-128.1 (1994), pp. 31–36 (cit. on p. 122).

[172]   D. Marušič and R. Scapellato. "Classifying vertex-transitive graphs whose order is a product of two primes". In: *Combinatorica* 14.2 (1994), pp. 187–201 (cit. on p. 122).

[173]   D. Marušič and R. Scapellato. "Permutation groups with conjugacy complete stabilizer". In: *Discrete Math.* 134 (1994), pp. 93–98 (cit. on p. 122).

[174]   D. Marušič and R. Scapellato. "Permutation groups, vertex-transitive digraphs and semi-regular automorphisms". In: *Europ. J. Combinatorics* 19 (1998), pp. 707–712 (cit. on p. 115).

[175]   D. Marušič, R. Scapellato, and N. Zagaglia Salvi. "A characterization of particular symmetric (0, 1) matrices". In: *Linear Algebra Appl.* 119 (1989), pp. 153–162 (cit. on pp. 102, 108).

[176]   D. Marušič, R. Scapellato, and N. Zagaglia Salvi. "Generalized Cayley graphs". In: *Discrete Math.* 102.3 (1992). URL: http://dx.doi.org/10.1016/0012-365X(92)90121-U, pp. 279–285 (cit. on pp. 103, 108, 119, 120).

[177]   D. Marušič, R. Scapellato, and B. Zgrablič. "On quasiprimitive pqr-graphs." In: *Algebra Colloq.* 2.4 (1995), pp. 295–314 (cit. on p. 116).

[178]   P. Maynard. "On Orbit Reconstruction Problems". PhD thesis. UEA, Norwich, 1996 (cit. on p. 169).

[179]   P. Maynard and J. Siemons. "On the reconstruction index of permutation groups: semiregular groups". In: *Aequationes Math.* 64.3 (2002), pp. 218–231 (cit. on p. 169).

[180]   P. Maynard and J. Siemons. "On the reconstruction index of permutation groups: general bounds". In: *Aequationes Math.* 70.3 (2005), pp. 225–239 (cit. on p. 169).

[181]   B. D. McKay and A. Piperno. "Practical graph isomorphism, {II}". In: *Journal of Symbolic Computation* 60 (2014). URL: www.sciencedirect.com/science/article/pii/S0747717113001193, pp. 94–112 (cit. on pp. 15, 17).

[182]   B. D. McKay and C. E. Praeger. "Vertex-transitive graphs that are not Cayley graphs I". In: *J. Austral. Math. Soc. (Ser. A)* 56 (1994), pp. 53–63 (cit. on p. 62).

[183]   B. D. McKay and C. E. Praeger. "Vertex-transitive graphs that are not Cayley graphs II". In: *J. Graph Theory* 22.4 (1996), pp. 321–324 (cit. on p. 62).

[184]   J. Meier. *Groups, Graphs and Trees: An Introduction to the Geometry of Infinite Groups*. Vol. 73. London Mathematical Society Student Texts. Cambridge, 2008 (cit. on p. 63).

[185]   J. Meng and M. Xu. "Automorphisms of groups and isomorphisms of Cayley digraphs". In: *Australasian J. Combin.* 12 (1995), pp. 93–100 (cit. on p. 91).

[186]   J. Milnor. "A note on curvature and finite groups". In: *J. Diff. Geom.* 2 (1968), pp. 1–7 (cit. on p. 63).

[187]   J. Milnor. "Growth of finitely generated solvable groups". In: *J. Diff. Geom.* 2 (1968), pp. 447–449 (cit. on p. 63).

[188] V. B. Mnukhin. "Reconstruction of k-orbits of a permutation group". In: *Math. Notes* 42 (1987), pp. 975–980 (cit. on p. 169).

[189] V. B. Mnukhin. "The k-orbit reconstruction and the orbit algebra". In: *Acta Applic. Math.* 29 (1992), pp. 83–117 (cit. on p. 169).

[190] V. B. Mnukhin. "The reconstruction of oriented necklaces". In: *J. Combin., Inf. & Sys. Sciences* 20.1–4 (1995), pp. 261–272 (cit. on p. 169).

[191] V. B. Mnukhin. "The k-orbit reconstruction for abelian and Hamiltonian groups". In: *Acta Applic. Math.* 52 (1998), pp. 149–162 (cit. on p. 169).

[192] R. Molina. "Correction of a proof on the ally-reconstruction number of a disconnected graph. Correction to: "The ally-reconstruction number of a disconnected graph" [Ars Combin. **28** (1989), 123–127; MR1039138 (90m:05094)] by W. J. Myrvold". In: *Ars Combin.* 40 (1995), pp. 59–64 (cit. on p. 143).

[193] R. Molina. "The edge reconstruction number of a disconnected graph". In: *J. Graph Theory* 19.3 (1995), pp. 375–384 (cit. on p. 143).

[194] A. Mowshowitz. "The group of a graph whose adjacency matrix has all distinct eigenvalues". In: *Proof Techniques in Graph Theory*. Ed. by F. Harary. Academic Press, 1969, pp. 109–110 (cit. on p. 38).

[195] A. Mowshowitz. "The adjacency matrix and the group of a graph". In: *New Directions in the Theory of Graphs*. Ed. by F. Harary. Academic Press, 1973 (cit. on p. 38).

[196] V. Müller. "Probabilistic reconstruction from subgraphs". In: *Comment. Math. Univ. Carolinae* 17 (1976), pp. 709–719 (cit. on pp. 38, 169).

[197] M. Muzychuk. "Ádám's conjecture is true in the square-free case". In: *J. Combin. Theory (Ser. A)* 72 (1995), pp. 118–134 (cit. on p. 122).

[198] M. Muzychuk. "On Ádám's conjecture for circulant graphs". In: *Discrete Math.* 176 (1997), pp. 285–298 (cit. on p. 122).

[199] M. Muzychuk and M. Klin. "On graphs with three eigenvalues". In: *Discrete Math.* 189.1-3 (1998), pp. 191–207 (cit. on p. 73).

[200] W. Myrvold. "Ally and Adversary Reconstruction Problems". PhD thesis. University of Waterloo, Ontario, Canada, 1988 (cit. on pp. 142, 144, 148).

[201] W. J. Myrvold. "The ally-reconstruction number of a disconnected graph". In: *Ars Combin.* 28 (1989), pp. 123–127 (cit. on p. 143).

[202] W. J. Myrvold. "The ally-reconstruction number of a tree with five or more vertices is three". In: *J. Graph Theory* 14 (1990), pp. 149–166 (cit. on pp. 144, 147).

[203] C. St. J. A. Nash-Williams. "The reconstruction problem". In: *Selected Topics in Graph Theory*. Ed. by L. W. Beineke and R. J. Wilson. Academic Press, 1978. Chap. 8 (cit. on pp. 123, 169).

[204] R. Nedela and M. Škoviera. "Which generalized Petersen graphs are Cayley graphs?" In: *J. Graph Theory* 19 (1995), pp. 1–11 (cit. on pp. 111, 122).

[205] E. D. Nering. *Linear Algebra and Matrix Theory*. John Wiley & Sons, 1970 (cit. on p. 77).

[206] J. Nešestřil and V. Rödl. "Products of graphs and their applications". In: Lecture Notes in Mathematics 1018. Ed. by J. W. Kennedy, M. Borowiechki and M. M. Syslo. Springer-Verlag, 1983, pp. 151–160 (cit. on p. 105).

[207] R. J. Nowakawski and D. F. Hall. "Associative products and their independence, domination and coloring numbers". In: *Discuss. Math. Graph Theory* 16.1 (1996), pp. 53–79 (cit. on p. 108).

[208] L. A. Nowitz and M. E. Watkins. "Graphical regular representations of non-abelian groups, I". In: *Canadian J. Math* XXIV.6 (1972), pp. 993–1008 (cit. on p. 91).

[209] L. A. Nowitz and M. E. Watkins. "Graphical regular representations of non-abelian groups, II". In: *Canadian J. Math* XXIV.6 (1972), pp. 1009–1018 (cit. on p. 91).

[210] V. Nýdl. "Some results concerning reconstruction conjecture". In: *Proceedings of the 12th winter school on abstract analysis (Srní, 1984)*. Suppl. 6. 1984, pp. 243–246 (cit. on p. 138).

[211] V. Nýdl. "A note on reconstructing of finite trees from small subtrees". In: *Acta Univ. Carolin. Math. Phys.* 31.2 (1990). 18th Winter School on Abstract Analysis (Srní, 1990), pp. 71–74 (cit. on p. 138).

[212] O. Ore. *Theory of Graphs*. American Mathematical Society, 1962 (cit. on p. 94).

[213] W. Pacco and R. Scapellato. "Digraphs having the same canonical double covering". In: *Discrete Math.* 173 (1997), pp. 291–296 (cit. on pp. 106, 108).

[214] M. Petersdorf and H. Sachs. "Spectrum und Automorphismengruppe eines Graphen". In: *Combinatorial Theory and Its Applications*. Vol. III. North-Holland, 1969, pp. 891–907 (cit. on p. 38).

[215] L. Porcu. "Sul raddoppio di un grafo". In: *Att. Ist. Lombardo (Rend. Sci.)* A.110 (1976), pp. 353–360 (cit. on p. 108).

[216] M. Pouzet. "Application de la notion de relation presquenchaînable au dńombrement des restrictions finies d'une relation". In: *Z. Math. Logik Grundl. Math.* 27 (1981), pp. 289–332 (cit. on p. 63).

[217] L. Pyber. "The edge-reconstruction of Hamiltonian graphs". In: *J. Graph Theory* 14 (1990), pp. 173–179 (cit. on p. 169).

[218] A. J. Radcliffe and A. D. Scott. "Reconstructing subsets of $\mathbb{Z}_n$". In: *J. Combin. Theory (Ser. A)* 83.2 (1998), pp. 169–187 (cit. on p. 169).

[219] A. J. Radcliffe and A. D. Scott. "Reconstructing subsets of reals". In: *Electron. J. Combin.* 1 (1999), Research Paper 20 (cit. on p. 169).

[220] S. Ramachandran. "N-reconstructibility of nonreconstructible digraphs". In: *Discrete Math.* 46.3 (1983), pp. 279–294 (cit. on p. 137).

[221] S. Ramachandran. "Reconstruction number for Ulam's conjecture". In: *Ars Combin.* 78 (2006), pp. 289–296 (cit. on p. 144).

[222] J. J. Rotman. *An Introduction to the Theory of Groups (4th Ed.)* Springer-Verlag, 1995 (cit. on pp. 7, 17).

[223] G. Sabidussi. "On the minimum order of graphs with a given automorphism group". In: *Monatsh. Math.* 63 (1959), pp. 124–127 (cit. on p. 62).

[224] G. Sabidussi. "Graph multiplication". In: *Math. Z.* 72 (1960), pp. 446–457 (cit. on p. 108).

[225] G. Sabidussi. "Vertex transitive graphs". In: *Monat. Math.* 68 (1964), pp. 426–438 (cit. on p. 122).

[226] H. Sachs. "Über Teiler, Faktoren und characterische Polynome von Graphen II". In: *Wiss. Z. Techn. Hosch. Ilmenau* 13 (1967), pp. 405–412 (cit. on p. 77).

[227] W. A. Stein et al. *Sage Mathematics Software (Version 6.3)*. www.sagemath .org. The Sage Development Team. 2014 (cit. on p. 15).

[228] R. Scapellato. "On *F*-geodetic graphs". In: *Discrete Math.* 80 (1990), pp. 313–325 (cit. on p. 38).

[229] R. Scapellato. "A characterization of bipartite graphs associated with BIB-designs with $\lambda = 1$". In: *Discrete Math.* 112 (1993), pp. 283–287 (cit. on p. 38).

[230] R. Scapellato. "Vertex-transitive graphs and digraphs". In: *Graph Symmetry: Algebraic Methods and Applications*. Ed. by G. Hahn and G. Sabidussi. Kluwer Acad. Publ., 1997, pp. 319–378 (cit. on pp. 106, 116, 122).

[231] A. J. Schwenk. "Spectral reconstruction problems". In: *Topics in Graph Theory*. Ed. by F. Harary. Vol. 328. Annals New York Academy of Sciences. New York Academy of Sciences, 1979, pp. 183–189 (cit. on p. 148).

[232] I. Sciriha. "Polynomial reconstruction and terminal vertices". In: *Linear Algebra Appl.* 356 (2002). Special issue on algebraic graph theory (Edinburgh, 2001), pp. 145–156 (cit. on p. 148).

[233] A. Seress. "On vertex-transitive, non-Cayley graphs of order *pqr*". In: *Discrete Math.* 182 (1998), pp. 279–292 (cit. on p. 62).

[234] Y. Shibata and Y. Kikuchi. "Graph products based on distance in graphs". In: *IEICE Trans. Fundamentals* E83-A.3 (2000), pp. 459–464 (cit. on p. 108).

[235] L. H. Soicher. "GRAPE: a system for computing with graphs and groups". In: *Groups and Computation*. Ed. by L. Finkelstein and W.M. Kantor. Vol. 11. DIMACS Series in Discrete Mathematics and Theoretical Computer Science. American Mathematical Society, 1993, pp. 287–291 (cit. on pp. 15, 17).

[236] T. Spence. *Strongly Regular Graphs on at Most 64 Vertices*. URL: www.maths .glaac.uk/~es/srgraphs.php (cit. on p. 78).

[237] P. K. Stockmeyer. "The falsity of the reconstruction conjecture for tournaments". In: *J. Graph Theory* 1 (1977), pp. 19–25 (cit. on p. 137).

[238] P. K. Stockmeyer. "Tilting at windmills or my quest for non-recon structible graphs". In: *Congressus Numerantium* 63 (1988), pp. 188–200 (cit. on p. 137).

[239] D. B. Surowski. "Stability of arc-transitive graphs". In: *J. Graph Theory* 38.2 (2001), pp. 95–110 (cit. on p. 108).

[240] D. B. Surowski. "Unexpected symmetries in unstable graphs". In: *J. Graph Theory* 38 (2001), pp. 95–110 (cit. on p. 108).

[241] D. B. Surowski. "Automorphism groups of certain unstable graphs". In: *Math. Slovaca* 53.3 (2003), pp. 215–232 (cit. on p. 108).

[242] R. G. Swan. "A simple proof of Rankin's campanological theorem". In: *Amer. Math. Monthly* 106 (1999), pp. 159–161 (cit. on p. 62).

[243] The GAP Group. *GAP—Groups, Algorithms and Programming, Version 4.1*. www-gap.dcs.st-and.ac.uk/ gap: Aachen, St. Andrews, 1999 (cit. on pp. 15, 17).

[244] R. M. Thomas. "Cayley graphs and group representations". In: *Math. Proc. Camb. Philos. Soc.* 103 (1988), pp. 385–387 (cit. on p. 62).

[245] C. Thomassen. "A characterisation of locally finite vertex-transitive graphs". In: *J. Combin. Theory (Ser. B)* 43 (1987), pp. 116–119 (cit. on p. 91).

[246] V. I. Trofimov. "Graphs with polynomial growth". In: *Math. USSR Sbornik.* 59 (1985), pp. 405–417 (cit. on p. 63).

[247]   V. I. Trofimov. "On the action of a group on a graph". In: *Acta. Appl. Math.* 29 (1992), pp. 161–170 (cit. on p. 63).

[248]   J. Turner. "Point-symmetric graphs with a prime number of points". In: *J. Combin. Theory* 3 (1967), pp. 136–145 (cit. on p. 122).

[249]   W. T. Tutte. "A family of cubical graphs". In: *Proc. Cambridge Philos. Soc.* 43 (1947), pp. 26–40 (cit. on p. 38).

[250]   W. T. Tutte. *Connectivity in Graphs.* Toronto University Press, 1966 (cit. on p. 38).

[251]   W. T. Tutte. "All the king's horses—a guide to reconstruction". In: *Graph Theory and Related Topics.* Ed. by J. A. Bondy and U. S. R. Murty. Academic Press, 1979 (cit. on p. 156).

[252]   J. H. van Lint and R. M. Wilson. *A Course in Combinatorics.* Cambridge University Press, 2001 (cit. on p. 148).

[253]   M. E. Watkins. "Connectivity of transitive graphs". In: *J. Combin. Theory* 8 (1970), pp. 23–29 (cit. on p. 60).

[254]   M. E. Watkins. "Ends and automorphisms of infinite graphs". In: *Graph Symmetry: Algebraic Methods and Its Applications.* Ed. by G. Hahn and G. Sabidussi. Kluwer Acad. Publ., 1997, pp. 379–414 (cit. on pp. 63, 77).

[255]   R. Weiss. "The non-existence of 8-transitive graphs". In: *Combinatorica* 1 (1983), pp. 309–311 (cit. on p. 38).

[256]   M. Welhan. "Reconstructing trees from two cards". In: *J. Graph Theory* 63 (2010), pp. 243–257 (cit. on p. 144).

[257]   D. B. West. *Introduction to Graph Theory.* Prentice-Hall, 1996 (cit. on pp. 3, 17).

[258]   H. Whitney. "Congruent graphs and the connectivity of graphs". In: *Amer. J. Math.* 54 (1932), pp. 150–168 (cit. on pp. 12, 139).

[259]   R. J. Wilson. *Introduction to Graph Theory.* Longman, 1997 (cit. on pp. 3, 17).

[260]   S. Wilson. "Unexpected symmetries in unstable graphs". In: *J. Combinatorial Theory* B.98 (2005), pp. 359–383 (cit. on p. 108).

[261]   D. Witte and J. Gallian. "A survey: Hamiltonian cycles in Cayley graphs". In: *Discrete Math.* 51 (1984), pp. 293–304 (cit. on p. 62).

[262]   J. A. Wolf. "Growth of finitely generated solvable groups and curvature of Riemannian manifolds". In: *J. Diff. Geom.* 2 (1968), pp. 421–446 (cit. on p. 63).

[263]   M. Y. Xu. "Automorphism groups and isomorphisms of Cayley digraphs". In: *Discrete Math.* 182 (1998), pp. 309–319 (cit. on p. 91).

[264]   Mingyao Xu and Shangjin Xu. "Symmetry properties of Cayley graphs of small valencies on the alternating group A_5". In: *Sci. China Ser. A* 47.4 (2004), pp. 593–604 (cit. on p. 91).

[265]   H. Yuan. "An eigenvector condition for reconstructibility". In: *J. Combin. Theory (Ser. B)* 32 (1982), pp. 245–256 (cit. on p. 148).

[266]   Y. Zhao. "On the edge reconstruction of graphs embedded in surfaces. III". In: *J. Combin. Theory (Ser. B)* 74 (1998), pp. 302–310 (cit. on p. 139).

[267]   A. A. Zykov. *Fundamentals of Graph Theory.* Translated from the Russian and edited by L. Boron, C. Christenson and B. Smith. Moscow, ID: BCS Associates, 1990 (cit. on pp. 99, 108).

List of Notations

185

$(\Gamma_1, Y_1) \equiv (\Gamma_2, Y_2)$, 3
$G \boxtimes H$, 97
GGP, 93
G^k, 93
$G \Diamond H$, 99
$\mathcal{G}(n, p)$, 26
GPG(n, k), 110
GRR, 79
$G \Box H$, 97
$G \propto H$, 97
$G \times H$, 94, 96

$I(u, v)$, 98
id, 93

$K(n, k)$, 66, 114
$\kappa(\mathcal{F}, G)$, 150
K_n, 8
$K_{p,q}$, 19

$L(G)$, 10
$L(\Gamma)$, 3
$\lambda_a^{\mathcal{H}}$, 3
λ_a, 3
$L^{\mathcal{H}}(\Gamma)$, 3
$loop$, 93

$m(D)$, 2
$m(G)$, 2

$n(D)$, 2
$n(G)$, 2
$N_i(u, v)$, 37

$o(1)$, 36
O_k, 66
O_{k+1}, 114

$\phi(G; x)$, 145
$\phi(G; x)$, 29
P_{k+1}, 2

$QG(Q, E)$, 117
Q_n, 37

$R(\Gamma)$, 3
$r(k, k)$, 36
$\rho(\Gamma, D)$, 166
$\text{rn}(G)$, 142
$(R \to T)$, 159
$(R \to_X T)$, 161

$\binom{S}{R}$, 158
$S(k, m)$, 149
$s(X)$, 154
$\mathcal{SD}(S)$, 158
$S_n^{(2)}$, 158
S_Y, 3

$[T]_R$, 160
$[T]_{R \backslash X}$, 161

uv, 1, 2

$V(G)$, 1

xTx^t / xx^t, 145

$\binom{Y}{2}$, 5, 158

$\mathbb{Z}_m^*$, 115
$\mathbb{Z}_n$, 6
$\mathbb{Z}_p^*$, 90

Index of Terms and Definitions

$\frac{1}{2}$-arc-transitive, 24

a.e., 26
Ádám's Conjecture, 91, 122
adjacency matrix of a graph, 29
adjacent, 1
adjacent to, 2
adversary reconstruction number of a graph, 142
affine group of transformations of a field, 121
ally reconstruction number of a graph, 142
almost every, 26
arc, 2
arc-transitive graph, 23
association schemes, 77
asymmetric graph, 18
automorphism
    colour preserving, 41
    of digraph, 7
    of graph, 7
    of structure, 160
    two-fold, 103

ball, 57

canonical double cover, 102
canonical double covering
    of a digraph, 106
card, 124
cartesian product, 94, 97
cartesian sum, 94
categorical product, 92
Cayley colour graph, 39
Cayley digraph, 44
Cayley graph, 45
    Hamiltonicity of, 53–54

character of abelian group, 55
character table, 55
characteristic polynomial deck, 145
characteristic polynomial of a graph, 29
circulant graph, 109, 116
class-edge-reconstruction number, 142
class-reconstruction number of a graph, 142
classification of vertex-transitive graphs of order pq, 116, 122
clique, 99
coherent configuration, 77
colour preserving automorphism, 41
commutator, 16
compatible structures, 157
complement of a graph, 1
complete bipartite graph, 19
complete graph, 8
complexity of graph isomorphism, 14, 139
conference graph, 75
connectedness preserving product, 96
connecting set, 45
coset graph, 49
cover of a graph, 149
cube, 37
cubic graph, 1
cutvertex, 128
cycle, 2
cyclic group, 6

deck of a graph, 124
degree, 1
degree of an edge, 137
degree-associated adversary reconstruction number of a graph, 143

187

Printed in the United States
by Baker & Taylor Publisher Services